The Best
of
Rational Science

Monk E. Mind

Other Books by Monk E. Mind
www.monkemind.com

ISBN-13: 978-1547219407
ISBN-10: 1547219408

Acknowledgments

Many thanks to friends of the Rational Scientific Method and Rope Hypothesis facebook groups from whose dedication to rational science I have drawn great inspiration and from valuable years of discussion much of this material has arisen.

Many, many thanks to Bill Gaede for a rational Scientific Method, Rope Hypothesis and Thread Theory, for his encouragement on this book, for many hours of science discussion, and for giving me the tools to understand the difference between fantasy and reality and how to determine the difference between possible and not possible.

As always, thank you, Mom and Dad. Without you I would not have been possible.

CONTENTS

Monk E. Mind

Chapter One - The Rational Scientific Method

Hypothesis, Theory, and Conclusion: A Rational Scientific Method of Inquiry

In science, a definition is a limitation or restriction on the use of a word. Scientific definitions are rational, non-contradictory, unambiguous terms that are consistently used and narrowly defined by the person who is making the hypothesis. We use adjectives to modify nouns (objects) and adverbs to qualify verbs (concepts). Science, in general, and physics in particular are about the physical... those things which have physical presence: what is real; things that exist.

To exist means to have shape and location, that is, an object with a location; something, somewhere. We visualize objects and we explain concepts. We do not explain objects - we point to them. We explain phenomena. The scientific method is based on hypothesis and theory. The conclusion is left to each individual. The hypothesis includes the statement of facts, the definitions of key terms, and the objects. The hypothesis describes the phenomena and illustrates the objects, defines the key terms, then makes the assumption(s). Assumptions are statements of the facts - not the facts themselves. Assumptions are neither true nor false. One does not define objects; one illustrates them. The theory explains the phenomena of the hypothesis. Everyone must decide for themselves.

Each individual concludes that the theory is either possible or not possible. Science is about explaining. Science in general, and physics in particular are about physically present objects.

Understanding the difference between objects and concepts allows one to make a rational conclusion about the key terms and the statement of facts at the hypothesis stage of the scientific method.

Proof is for math. Science never proves. Science is about physical reality. Math describes abstract dynamic concepts, whereas science illustrates static physical objects, and explains phenomena. Math is NOT the language of science, illustration is.

A hypothesis stands on its own. It does not matter who agrees. The hypothesis should illustrate the objects, define the key terms, and present the statement of the facts, the assumptions. The theory would then explain the phenomena of the hypothesis. There is no correct or incorrect hypothesis - it is an assumption. It is either rational or not. If it is rational, we accept the assumption(s) of the hypothesis. Predictions and observations are opinions and are extra-scientific.

Hypotheses are assumptions, and theories explain the hypotheses. We form a conclusion that the theory is either possible or not possible. This is why in science it is crucial to understand the difference between objects and concepts, nouns and verbs, adjectives and adverbs, hypotheses and theories. We can say: I see a field of corn. The corn stalks wave in the wind.

I have a dust particle in my eye. BUT...fields, waves, and mass-less particles are concepts in math which do not exist in physical reality, therefore, should not be presented in the hypothesis. We describe or illustrate objects in the hypothesis.

We explain concepts in the theory. We never explain objects; we describe them, illustrate them, or point to them.

"Insofar as mathematics is exact, it does not apply to reality; and insofar as mathematics applies to reality, it is not exact." – Einstein See, I told you sometimes Einy made sense!

The mathematical physicist uses ambiguous or contradictory terms inconsistently. He or she confuses objects with concepts, nouns with verbs, adverbs with adjectives, and hypotheses with theories. Reality does not depend on human perception or observation. It is because the human senses are limited and flawed that science must be as objective. The scientific method should be observer independent as much as possible. A rational key term never invokes an observer. Although our senses are limited, there is no limit to our intellect. One must apply rationality, reasoning, and critical thought at the conceptual stage in the hypothesis. Precision is precious. Defining key terms is critically important. Understanding the difference between concepts and objects is essential in dealing with science. We make this clear with our definitions. In science, one must be able to visualize the concrete object. Objects must be illustrated in the hypothesis. The objects are the actors, the key terms make clear the meaning of the script, and the statement of facts sets the initial scene for the theory. The dynamic concepts in the theory are describing the phenomena of the hypothesis. The hypothesis is a photo (static), the theory is a movie (dynamic).

Each person takes away their own conclusion as to whether or not the story was possible.

Most important are the key terms, and these words have meaning as defined by the theorist. In science, one can only use objects that can be illustrated in the hypothesis. If it cannot be illustrated or visualized, then it is not real and has no physical presence. What is not physical has no place in science.

Science, especially physics, is conceptual. Technology, which is mostly trial & error, is empirical.

Planes that fly, microwaves that heat, and GPS devices that measure your position work primarily through trial and error because of technology...not because the theories that they are supposedly founded upon are "correct."

The problem lies in the confusion between objects and concepts. There is no good way to discuss General or Special Relativity, Quantum Mechanics, or String Theory until point, line, and plane can be defined and understood. Math attempts to describe dynamic concepts by moving numbers. Physics is about reality.

What exists, physically present objects with location, are made up of matter. These are static and can be photographed or illustrated. But we must be able to define what 'exist' means.

Universe: matter (atoms) and space (nothing)

Concept: the relationship between two or more objects

Object: that which has shape

Space: that which does not have shape

Exist: matter + location

Location: the set of static distances to all other objects

Motion: object + 2 or more locations

Theoretical physics, Newtonian physics, ToR, and QM don't explain anything, they describe. These theories predict or describe, but do not explain. It is not interesting that Newton tells me an apple falls at 9.8 meters per ft per second per second. I want to know why. I can point at an apple and say, "Look it is falling fast." So what? What is the physical medium that attracts objects to each other? That is the question for science. Math 'predicts' how fast something falls to the ground, but it says nothing about why it falls.

"Since the mathematicians have invaded the theory of relativity, I do not understand it myself anymore."—Albert Einstein

Ptolemy 'predicted' to a high degree of accuracy the position of the planets in the solar system, but he had the earth in the center. That does not help explain why the planets orbit in elliptical paths and don't fly out into space.

What about these 'predictions'? If I observe an apple fall a few times and measure the speed and distance traveled, I can 'predict' how fast an apple falls. What does that tell me? It does not tell me when an apple is going to fall. Now THAT would be a real prediction. Something that already happened, a consummated event, is described and should then be explained.

Something that we have observed happen repeatedly can lead us to think that there is a high degree of probability that it will happen that way again. But that is not really a prediction - it's an educated guess.

Belief, truth, evidence, and proof are not part of the scientific method; it is observer-independent. Experiments and observation are extra-scientific. Science, especially theoretical physics, is conceptual. Technology, mostly trial & error, is empirical. Here's the root of the problem with the

currently taught scientific method: It all revolves around simple misunderstandings of basic physical reality brought on by the inability to determine the difference between an object and a concept, and the inability to precisely and consistently define terms upon which a theory depends.

At the root of the Relativity and Quantum Mechanics problem is Euclidean geometry. Because the point, the line, and the plane are not defined, or are defined ambiguously using abstract concepts instead of objects, they do not represent actual physical reality! A rather shaky basis on which to form the physical 'laws' of the universe.

Rational Scientific Method :

Hypothesis: defines our key terms and makes a statement of the facts, the assumptions. We assume in the hypothesis stage. If the assumptions are rational, then we can proceed to the theory.

The objects of the hypothesis are described or illustrated, a photograph-static.

Theory: explains the hypothesis; phenomena such as motion or process, a movie-dynamic.

Conclusion: possible or not possible? Everyone decides for themselves.

If the key terms of the hypothesis are ambiguous, circular, synonymous, or contradictory, then the theorist should throw out the hypothesis, or present precise, rational definitions of key terms upon which the hypothesis depends.

The theory is where we present a 'movie' or series of illustrations of the phenomena, or process, involved in explaining the hypothesis. Then, and only then, can we form our conclusion.

If we conclude the theory is irrational, and therefore not possible, we throw the theory out.

If we conclude that the theory is possible, then we publish a paper, or stand around the water cooler telling people about it, or simply move on to the next thing on our agenda. If we conclude that the theory is possible, but does not provide the complete explanation, we form another hypothesis based upon the theory and build upon it. The flat earth becomes the round earth, which becomes the oblate spheroid... Once the theory is presented, science is done!

The conclusion is left up to each individual: Possible or NOT Possible

Chapter Two - Scientist, Science, & the Scientific Method

Recently, I was reading to my wife from some correspondence between me and an astrophysicist, who referred to me as a scientist.

My wife made a guttural sound and rolled her eyes at the mention of me as a scientist. She has multiple degrees, and took exception to the notion that I was a scientist with my single degree in electrical engineering and no other formal training or affiliation in science.

I explained that the man knew my background, and although he certainly had the credentials she was looking for, both he and I had a different idea than she of what being a scientist is all about.

To many, the word scientist conjures up images of men and women wearing white lab coats with nerd pockets filled with pens, magnifier glasses, and slide rules.

Let me ask you. Which schools does a scientist attend, and how many diplomas and certificates of achievement does it take to qualify as a scientist? How many degrees does it take to be a scientist? Which laboratories or corporations does a scientist work in?

What is science, and what does it mean to be a scientist?

From MonkEpedia:

science: the systematic study of reality

Discussion: Science offers objective explanations for natural phenomena using the Rational Scientific Method of inquiry

This from FatFist:

"Science: the study of reality (existence) for the purposes of accumulating a collection of rational explanations (i.e. Theories) for natural phenomena using the Scientific Method (Hypothesis + Theory)."

This from Bill Gaede:

"science: the body of papers that follow the scientific method which man has accumulated over the years."

So what then is a scientist? If science explains, then it is the scientist who does the explaining.

To quote from Bill's book, "Why God Doesn't Exist":

"In my humble opinion, the definition of science boils down to a choice between the investigative and publishing aspects. These activities entail two different sets of skills. I will refer to the investigator as a detective and the individual who makes a theory public as the prosecutor. The detective is a researcher, an inventor, an engineer, a person who usually develops technology or solves technical problems. This is the hero of the establishment, but again, in my opinion, has little if anything to do with science. It is the prosecutor, the individual who communicates his findings to the world, who has to do with science. Science has to do with communication, and not with discovery."

Science is the explanation, Scientists are the 'explainers' and the Scientific method is how scientists arrive at the explanations.

I would like to suggest to you, that as you talk to others, as you debate on the internet, and as you write about rational science, the Rational Scientific Method, and refute the ridiculous proposals of the mathemagicians, it is YOU who are explaining reality, and it is YOU who are the true scientists.

Chapter Three - The Scientific Method? For Dummies!

What if I can't observe it?

Let's take a look at what is being passed off to children as the scientific method of inquiry.

Some of the following is taken from Science Made Simple.

Science Definition

"The word science comes from the Latin "scientia," meaning knowledge."

We are told that science refers to a method of obtaining knowledge using observation and experimentation to explain naturally occurring phenomena. What other kinds are there?

We are also told that the purpose of science is to produce "useful models of reality." Yet nowhere is "reality" defined.

According to this particular author the steps of the Scientific Method are:

"Observation: ...it's important to use as many sources as you can find."

IOW, get as many opinions that you can first. This is what they are teaching our youth. Don't think for yourself, depend on Polly-parroting the 'experts.'

"Next is hypothesis: This word basically means 'a possible solution to a problem, based on knowledge and research.' The hypothesis is a simple statement that defines what you think the outcome of your experiment will be."

This person confuses religion with science. A hypothesis is a statement about what the theorist thinks the problem is, not about knowledge and not about research! What does "knowledge" have to do with the hypothesis, or what is? If I don't "know" the moon is there, does it cease to exist? Guess I better consult with the All

Knowing! Experimentation is extra-scientific. But... Don't worry; later you can change your hypothesis, if the results don't match.

"PREDICTION: The hypothesis is your general statement of how you think the scientific phenomenon in question works. Your prediction lets you get specific -- how will you demonstrate that your hypothesis is true? The experiment that you will design is done to test the prediction."

Phenomena? We haven't even got to the theory yet! They confuse hypothesis and theory. The phenomena are explained in the theory. Experiments are extra-scientific, and true/false have to do with religion and philosophy, not science. Science is about explaining, and those explanations have to be rational. If they are not rational, it is not science!

"EXPERIMENT: This is the part of the scientific method that tests your hypothesis. An experiment is a tool that you design to find out if your ideas about your topic are right or wrong."

IOW, does the experiment verify our religious beliefs? Right and wrong are not part of science, they are part of religion!

AND FINALLY...

"CONCLUSION: The final step in the scientific method is the conclusion. "

The conclusion is NOT part of the scientific method of inquiry!

This is what is expected to get a blue ribbon at the science fair (all in preparation for the even bigger prize, along with a tiara and a bouquet of roses) a piece of paper saying how smart you are, adoration, pats on the back from your peers, and possibly grant money....

This best explains the problem with ...The (so-called) Scientific Method

1st we look for truth...We use deduction...Then add a little induction...

And we get...

Circular Reasoning: "The two methods (of induction and deduction) should be seen as a cycle, with inductive reasoning generating a theory, with deduction and experimentation validating or falsifying the theory. This, in turn, leads to inductive enhancements of the theory and more testing"...and chasing our tails...

Yep...that says it all. We use our observations to test for truth and chase our tails around until we dig a rut so deep, we can't get out!

Or, we could use the Rational Scientific Method and spot circular reasoning right at the hypothesis stage, erase the whiteboard and head to the bar. Or, start all over with a rational hypothesis.

We talked about rational thinking a bit in "the Rational Thinking Test."

But...What is a rational Scientific Method? Again? Yes, it is very, very important!

Science is about explaining phenomena using objects. All phenomena are the result of surface to surface contact between objects. So we must first hypothesize the objects.

Hypothesis contains three main components:

1) The exhibits: or, actors; objects. This requires photographs, illustrations, mock-ups, or sculptures so that everyone can visualize exactly what objects are involved.

2) The Key Terms: the theorist is responsible for using precise terms that HE defines so everyone understands what it is he is talking about! A rational definition (or limit on the use of a term) is unambiguous, non-synonymous, non-circular, non-contradictory, and can be used consistently throughout a presentation.

3) A statement of the facts (not to be confused with the facts): This sets the scene for the explanation in the theory.

Theory: The theory is an explanation of the phenomena using objects.

Conclusion: After the theorist's presentation with his hypothesis and theory, science is done! The conference attendees must decide for themselves if what was proposed in the theory is possible or not possible.

Observation and experimentation are extra-scientific. We leave this to the technicians and engineers in the Technology Department down the hall.

For questions and discussions about the Rational Scientific Method, join our Face book group 'Rational Scientific Method.'

Chapter Four - Hypothesis, Theory, Conclusion

Anyone following my writings, ramblings and rants, has heard me say a million times that persons confuse objects with concepts, verbs with nouns, hypothesis with theory, and fantasy with reality.

There is also confusion about facts and opinions. They are one and the same! The confusion between facts & opinions is directly linked to the ideas of true/false, fact, and evidence. This manifests itself in the scientific method with confusion between hypothesis & theory and it all originates with the confusion between objects and concepts.

I have covered truth and authority in Rational Science Vol. I. but in short Truth = Opinion, and there are no authorities.

In this chapter I wish to take a look at hypothesis and theory and show how not understanding the difference between objects and concepts leads to confusion in the scientific method and how true, false, proof and evidence are linked to observation and experimentation.

Recently I was told I was forcing my opinions on others. Of course, I didn't see it that way as I was just trying to show the person their opinion was irrational. One needs to understand that there is a difference between rational opinions and irrational opinions. Everyone has opinions and it is well that they do. I merely wanted to point out that irrational opinions don't help anyone. This person was a perceived authority (a lawyer) speaking about truth and absolutes (an irrational proposal) in a blog. Others were offering their own differing definitions. His proposition began with a contradictory (irrational) definition for absolute and no definition for truth. The lack of definitions, or, irrational definitions, leads to irrational assumptions and opinions.

Since he was a lawyer, I tried to explain that facts are opinions from perceived authority. The prosecutor or defense attorney presents their evidence (opinions), facts (opinions of experts or authorities) and witnesses (fallible observations) to 'prove' their case. A higher court may have different opinions, and based on the same 'proof' overturn the lower court's ruling.

This is an accepted method in the courts, but true, false, facts, and evidence are all opinions which have no place in the hypothesis or theory of a scientific method of inquiry.

These concepts are useful for persuasion in a court of law and also for convincing in philosophy and religion. Opinions belong in the conclusion stage of the Rational Scientific Method (RSM). Science is not concerned with what the individual concludes. Their opinion is theirs alone

A scientific conclusion can only be possible or NOT possible. The accused person's fingerprints may be on the murder weapon, but that does not mean he or she killed the victim.

Where do the concepts true, false, proof and fact come from? They come from individuals who rely on observation, experimentation and testing.

Everyone who has ever played the Gossip Game understands how unreliable observation is. Anyone who has seen a mirage, or had a vision or dream, understands the human senses are not always to be depended upon. Anyone who has worked in the court system understands that often the least reliable evidence is a witness.

We can thank Popper for bringing falsifiability into the scientific method, as if the idea of truth was not enough! This individual also said this:

"One should never get involved in verbal questions or questions of meaning, and never get interested in words. If challenged by the question of whether a word one uses really means this or perhaps that, then one should say: "I don't know and I'm not interested in meanings......one should never quarrel about words, and never get involved in questions of terminology. One should always keep away from discussing concepts" - Karl Popper

Whether something is true or false, as already stated, is a matter of opinion and has no place in the scientific method. In the Rational Scientific Method we define our Key Terms in the hypothesis stage and clearly describe or illustrate the objects. Then we can understand if our assumptions are rational.

There are consequences to irrational assumptions. For example, the idea of matter and energy equivalence is a result of the irrational assumption that matter can be created and that energy is some "thing." This is because persons do not understand that matter and motion are eternal, and energy is not rationally defined. This sort of proposal would not make it past the hypothesis stage of the RSM where the difference between the objects and concepts is made clear.

HYPOTHESIS

Looking through the various dictionary definitions one will read something like the following which illustrates what I am talking about very well:

Wikipedia: "A hypothesis is a proposed explanation for a phenomenon. For a hypothesis to be a scientific hypothesis, the scientific method requires that one can test it. Scientists generally base scientific hypotheses on previous observations that cannot satisfactorily be explained with the available scientific theories. Even though the words "hypothesis" and "theory" are often used synonymously, a scientific hypothesis is not the same as a scientific theory. A scientific hypothesis is a proposed explanation of a phenomenon which still has to be rigorously tested. In contrast, a scientific theory has undergone extensive testing and is generally accepted to be the accurate explanation behind an observation."

Note that the hypothesis stage here is an explanation (which really belongs in the theory) and involves observation and experimentation which is extra-scientific. Without objects there are no phenomena. Where are the objects?

Observation is what lead us to the inquiry, and is not part of the Rational Scientific Method. RSM is purely conceptual.

Free dictionary:

1. A tentative explanation for an observation, phenomenon, or scientific problem that can be tested by further investigation.

2. Something taken to be true for the purpose of argument or investigation; an assumption.

Wolfram:
A hypothesis is a proposition that is consistent with known data, but has been neither verified nor shown to be false.

Note the key terms here are: Testing, observation, true and false. One can see the scientific method as it is still being taught to school children in the chapter entitled, "Scientific Method? for Dummies!"

MonkEpedia:
The hypothesis in the RSM can be defined as the first step of the scientific method of inquiry, where an assumption about a phenomena is proposed for explanation by the second step, a theory, and is comprised of the following:

The objects

The Key Terms

The assumption (statement of the 'facts')

The objects are described, photographed, illustrated, or presented as a sculpture or mock-up, the Key Terms are defined in precise unambiguous, non-circular, non-contradictory and non-synonymous terms; the assumption(s) are understood to be neither right or wrong, true or false, correct or incorrect. They are assumptions used for purposes of explanation. If the first two items of the hypothesis are present, the assumption can be understood to be rational or not rational. If rational, precede to the theory.

THEORY

Merriam Webster:

"a plausible or scientifically acceptable general principle or body of principles offered to explain phenomena"

I love Webster gives the example "wave theory of light" (which is contradicted by the particle theory of light).

Note: See the chapters on light, Does it Travel Rectilinearly or Curvilinearly? And the chapters on distance.

Light is also covered in the chapter on the Particle Wave Paradox in the book Rational Science Vol. I.

Wolfram:
1 a well-substantiated explanation of some aspect of the natural world; an organized system of accepted knowledge that applies in a variety of circumstances to explain a specific set of phenomena

2 a tentative insight into the natural world; a concept that is not yet verified but that if true would explain certain facts or phenomena

Note here that the terms knowledge and truth are used, and something called insight is mentioned along with facts. Who substantiates the explanation? Whose insight and knowledge is accepted? What truth and what facts are part of the theory? Do we take a vote and determine this by a show of hands, or do we appeal to a higher authority?

MonkEpedia:
In the Rational Scientific Method of Inquiry the theory is the second step following the hypothesis where the phenomena of the hypothesis are explained.

The conclusion is yours to make. Based on the first definition for theory above, we can see that the conclusion has already been decided for you, as it is "well substantiated and accepted knowledge."

In short, when one rationally defines their terms everyone can understand what they are talking about (whether an object or a concept). When the objects are described or illustrated everyone can relate to the assumptions, and with the theory begin to understand the explanation of the phenomena.

When the Rational Scientific Method is not followed, we end up with fantasy, not reality.

Chapter Five - Science & Technology - Conceptual & Empirical

As covered in the chapter on the Rational Scientific Method, this author considers science as purely conceptual.

What I mean when I use the words science, scientist and the scientific method is defined in the chapter, "Scientist, Science & The Scientific Method."

Evidence, proof and facts are not considered scientific. They are not part of the Rational Scientific Method of inquiry because they are dependent upon the limited sensory system of man and are therefore considered opinion. Opinions are fine, and as the saying (partly) goes, "everyone has one." We reserve our opinions for after the hypothesis and theory is presented and it is called the conclusion (which is ours alone).

What is technology? We often see the phrase, "Science & Technology." Technology is empirical, evidence based, and mostly trial and error. The difference is often debated. A general consensus might be that science and technology are interdependent but separate endeavors.

From Ask.com we find the following:

"Science is a way of practicing knowledge, as well as the knowledge itself, whereas technology is the application of science, particularly to industrial or commercial objectives. Technology can also be defined as the scientific methods and materials used to achieve industrial objectives."

Before we continue, let's look at some common definitions.

According to Dictionary.reference.com, science is "a branch of knowledge or study dealing with a body of facts or truths systematically arranged and showing the operation of general laws: the mathematical sciences. 2. systematic knowledge of the physical or material world gained through observation and experimentation."

Knowledge, facts, and truth are the opinions of man, and as such, have no place in science or technology. That is part of religion and philosophy. Science explains and technology builds.

One would think that if science and technology are interdependent yet separate that there would be a friendly relationship and a clear division of labor between the two. Yet this is not the case. Scientists, engineers, social scientists, philosophers, historians, policy makers and the public all have self interests. Because of this, government and other agencies, like NASA, have Science and Technology departments just to help solve related issues and coordinate resources and development across the disciplinary boundaries.

The rivalry is one issue and the focus has been mostly on that. Just take a look at all the scholarly articles devoted to it. As an electrical engineer, and later a software and then hardware engineer, it was obvious to me who did the work and who got the credit.

Wiki Answers tell us about the relationship between science and technology:

"Science discovers fundamental information about how the universe works. Technology is the practical application of that information, or knowledge. A computer is an example of technology; in order to invent one, it is necessary to know a lot of fundamental science. Science sets the stage for technology, which produces useful devices. There would be no laptops without the fundamental discoveries of science."

We hear all the time that science is responsible for this and that. I've been told that if not for Quantum Mechanics the transistor would not have been invented. If not for relativity GPS would not have been possible and so forth. This is not the case as we shall see.

Many great discoveries had nothing to do with the 'science' behind it.

"Chance favors the prepared mind." - Louis Pasteur

Are we to believe that serendipity is part of the scientific method? We are told that it is:

"Serendipity means a "happy accident" or "pleasant surprise"; a fortunate mistake. Specifically, the accident of finding something good or useful while not specifically searching for it. ..Indeed, the scientific method, and the scientists themselves, can be prepared in many other ways to harness luck and make discoveries." - WIKI

Charles Goodyear accidentally spilled a mixture of rubber, sulfur, and lead onto a hot stove creating vulcanized rubber.

The engineer Wilson Greatbatch used the wrong value resistor and the circuit he was working on pulsed like a heart beat giving him the idea for the pace maker.

Alfred Nobel accidentally discovered dynamite when he dropped Nitroglycerin in sawdust. The sawdust soaked it up, stabilizing it and making it useful as dynamite.

Alexander Fleming discovered Penicillin a day after failing to clean up his work station.

Benedictus dropped a flask that had contained a liquid plastic that had evaporated. The flask didn't shatter and safety glass was the result.

While trying to make artificial quinine, Perkins made the first synthetic dye superior to any of the natural dyes available at the time. Chemistry quickly became a money making enterprise. Later a German bacteriologist by the name of Paul Ehrlich, used the dyes in a different manner. He used them in immunology and chemotherapy.

Potato chips were an accident as were popsicles, ice cream cones, and Coca Cola. Smart dust (silicon chips) used for sensors was an accident as well.

Saccharin was discovered because a chemist didn't wash his hands and chemical got on his wife's dinner rolls.

Percy Spencer was using a vacuum tube and aimed it at various items in the lab and accidentally melted a chocolate bar in his pocket. The beginning of the first microwave oven!

Wilhelm Roentgen was tinkering with a device when he noticed a fluorescent light flickering. He started putting various objects in front of it and discovered he could see the bones of his hands. The X-ray machine had its beginning! Later, when experimenting with X-rays, Becquerel accidentally exposed a photographic plate with a uranium rock and then with the help of the Curies discovered radioactivity.

Chemist Leo Hendrik Baekeland, trying to make a shellac alternative, produced a material in one of his experiments called Bakelite. He was going to use it to make phonograph records but found it could be used for many other things. Plastic is derived from it today.

Lysergic acid was absorbed through the skin of Albert Hofmann and he got a buzz. Timothy Leary was happy about that because LSD came about because of it!

Phizer discovered that a drug they were using in a clinical trial for heart conditions, although useless for that purpose, was great for erectile dysfunction. Viagra was born!

Smallpox vaccination, clinical use of insulin for diabetes, and the Pap smear...all a result of serendipity. As were post it notes, Cellophane, and Velcro. So were Play-Doh, Stainless Steel, the Ink Jet Printer, and Vaseline ...all a result of serendipity.

"The seeds of great discoveries are constantly floating around us, but they only take root in minds well prepared to receive them." - Joseph Henry

Thank God for open minds and thank God for Viagra and Vaseline!

While these are all good examples of accidents, luck and happenstance meeting up with open minded, observant men and women, it points to something very important. It points to the difference between science and technology, conceptualizing and observing.

The Free Dictionary tells us that conceptual is relating to mental conception and gives a use as "conceptual discussions that antedated development of the new product."

The same source defines empirical as: "Relying on or derived from observation or experiment: empirical results that supported the hypothesis." The second definition is revealing: "Guided by practical experience and not theory, especially in medicine."

Does science depend on accidents, happenstance or serendipity? Does empiricism depend on theory? No!

Rational Science takes this a step further. Empiricism is extra-scientific. That is, it is NOT part of the Rational Scientific Method of inquiry. Why? Because the human sensory system is limited but the ability to conceive is not.

It is technology with its empiricism, experimentation, and its trial and error that is responsible for our so-called scientific advancements. Science played a very minor role.

The mathematical theorist, for instance, may believe that their calculations confirm, and GPS proves, Relativity. They therefore take credit for our GPS system. As discussed in Distance Part Two, the calculus only confirms that cesium atoms are being stressed differently as the satellite orbits the earth than the cesium atoms in the clocks at the ground stations. The atoms were stressed. Time was not warped as proposed by the ridiculous theory of relativity!

The transistor was discovered in Bell Labs by engineers tinkering with components and not because of some theory or theories relating to the ridiculous notions of Quantum Mechanics!

The purpose of science is to explain and that depends on conceptualization. The usefulness of technology is in designing and building. This is accomplished by experimentation, testing, and trial and error which depend on observation.

Clearly, science and technology are two different things altogether. The problem with the modern scientific method is that they confuse the two. Scientists today confuse nouns with verbs, concepts with objects, hypothesis with theory, and science with technology.

Chapter Six - Experiments: Are They Part of the Scientific Method?

No, Kiera, they're not! As discussed in "Science is Conceptual, Technology is Empirical," experimentation is observation based and therefore extra-scientific.

A lot of money and resources will be saved when this is understood by the scientific community and a corresponding cooperative division of labor is established between science and technology.

In "Proof is for Alcohol," we learned that Aristotle conceived of a spherical earth long before Eratosthenes "predicted" it with math, and Magellan "proved" it by circumnavigating the globe.

In this chapter, let's take a look at experimentation and why it can go terribly wrong.

So what is being taught as the scientific method? This is it in a nutshell (from slideshare.net):

Research the question

Form a hypothesis

Conduct an experiment

Analyze the data

Draw conclusion

Communicate results

"Research the question" means see how other people have answered it. "Form a hypothesis" means repeat what you have read or heard about the question. "Conduct an experiment" means design a method that describes the hypothesis. "Analyze the data" means use observation to confirm your hypothesis. "Draw conclusion" means decide whether your hypothesis is right or wrong. "Communicate results" means to tell others what your conclusion   is.

This, from Science Projects.com

"The Job of the Scientist is to study the surrounding world and explain why the world is the way that it is. "

Good so far! Science should be about explaining.

"The way that this is carried out is by experimentation. The methods for producing experiments comprise what is called THE SCIENTIFIC METHOD."

Hmmm. Not so good. Observation is what brought us to this point. We saw something interesting and we want to understand the phenomena. It's circular to use observation to "explain" observation. Not only that, but our senses are flawed and limited as discussed in the Rational Scientific Method. Also, when we look closer we discover that experiments aren't explaining anything anyways, they are describing.

"When preparing to do research, a scientist must form a hypothesis, which is an educated guess about a particular problem or idea, and then work to support it and prove that it is correct, or refute it and prove that it is wrong.

"Whether the scientist is right or wrong is not as important as whether he or she sets up an experiment that can be repeated by other scientists, who expect to reach the same conclusion."- answers.com

We are told that the idea is to understand cause and effect using a controlled experiment which utilizes controls and variables. But understanding does NOT come from guessing, experimentation, right, wrong, or proof. Understanding comes when one can conceptualize the objects and rationally explain the phenomena. Experiments describe. Descriptions don't explain anything. We describe objects, we explain phenomena. We point at an apple, describe it, or offer a photograph of it. We attempt to explain WHY the apple fell onto Newton's head. We don't describe it falling at so many meters per second squared.

A kindergarten child understands that if the apple falls from the tree it moves real fast and lands on the ground. BUT WHY? Why doesn't the apple fall up into the sky? This is what everyone really wants to understand.

Who cares how many persons can repeat an experiment if we don't actually understand anything.

If you research using our dear friend Google "experiments gone wrong" you will generally find the type of accidental discoveries discussed in a previous chapter. The good results of bad science.

If you have been around the interwebs for any time, you have likely run across the many charts showing the correlation between the number of pirates and global temperature. Of course this is done in fun to illustrate the point that "correlation does not imply causation."

FAILED EXPERIMENTS

General R.G. Dyrenforth, the concussionist, represents another such example and brings it closer to home for us Texas boys... er ...MonkEs. Although not a General, or a Commissioner of Patents, as he claimed, he was able to convince many people into believing one could blow up explosives in the air and cause it to rain.

Others had believed that rain followed artillery. Plutarch, Napoleon, and Edward Powers erroneously believed this as well. In 1871 Powers wrote a book about it entitled "War and the Weather." He even convinced the US government into paying him $2,000 to make it rain. Dyrenforth was tasked with the job.

It's a funny part of Texas history, where experiments conducted in the heat of a Texas summer went comically wrong. Trainloads of dynamite and gunpowder were sent to Midland.

The "General" told New York Times reporters that it "is a matter of cold fact, based on my experiments, I know that rain can be produced."

Newspapers such as the Chicago Tribune, New York Sun, and the Washington Post, wrote articles about man's will to control nature for his own purposes. Most of these reporters didn't even go to Texas. However, Texas Farm and Ranch, and Farm Implement News reporters were there to witness the experiment. Canons shot explosives into the air, and kites and balloons carried various explosives up in to the wild blue yonder to be blown away by strong Texas winds and scattered around exploding at the wrong time and lighting all kinds of things on fire.

Scientific American magazine later said the experiments were "an expensive farce."

Of course, Powers never considered the probability of rain for any given location when there wasn't a battle, and the fraudulent General only succeeded when Mother Nature decided she was going to provide rain... in spite of his silly antics. People apparently had enough of him after he blew out some windows in a San Antonio Hotel and wiped out a nearby mesquite tree.

Chernobyl was a failed emergency shutdown experiment.

In 1962, Tusko the elephant was given LSD (3000 times the human dose) just to see what would happen. They just guessed this amount even though an elephant is about 90 times the weight of a human. The elephant died. The "scientists" told Science magazine, "It appears that the elephant is highly sensitive to the effects of LSD."

Look up the Monster Study where children were made to stutter just to test the idea of positive and negative reinforcement.

Check out the experiment by Psychologist Winthrop Kellogg and his wife. They raised their newborn son David along side a chimp named Guo until they saw that their son was more like a chimp then the chimp was like a human.

There are thousands of examples, but perhaps the most famous is the one by B.F. Skinner who raised his baby in a box. This one, however, went exactly as expected. Still according to the public, when they found out, it was terribly wrong.

Keep those psychologists away from me please!

Here's an experiment gone way wrong. A Florida teenager is charged with a felony because of her "failed" experiment. She was in a science class mixing various household chemicals when the Eight oz. water bottle exploded (no one was injured). Kiera Wilmont was taken away in handcuffs and expelled from her school.

What really went wrong? Was it the curious girl's experiment? Was it the fact that she was unsupervised in the science lab? I submit that nothing really went wrong based on what is being taught in these science labs. The problem is when science classes teach the scientific method involves experimentation.

Why are these failed "scientific experiments?" Because experiments are not part of the scientific method! They are part of technology's trial and error.

Is this science?

When Thomas Edison was interviewed by a young reporter who boldly asked Mr. Edison if he felt like a failure, and if he thought he should just give up by now, perplexed, Edison replied, "Young man, why would I feel like a failure? And why would I ever give up? I now know definitively over 9,000 ways that an electric light bulb will not work. Success is almost in my grasp." And shortly after that, and over 10,000 attempts, Edison invented the light bulb.

No, this was clearly trial and error. Eventually a fantastic result, but not scientific! Today scientists are still unable to explain electricity or light.

When persons believe science is based upon observation and experimentation, and true and false, this is the sort of thing that they may end up believing.

"The Real Story Behind America's UFO Connection and Area 51 (from The Mind of James Donahue)

"There is a story involving the U. S. Military's mysterious Area 51 in Nevada that is so incredible it makes seasoned psychics and remote viewers turn pale when they look at it."

Not if one dispensed with ridiculous notions of psychics and remote viewing which has yet to be rationally explained. Scientific method based on observation, government secrets and UFO reports abounding, is it any wonder the gullible public falls for this sort of nonsense?

"Aliens are real. They exist in a parallel universe that some people call the astral plane, or sub-space. It exists all around us. The aliens are mostly beings of light to us. The best way to get a good look at them is to learn how to leave our bodies and enter the parallel universe where they live."

Parallel universes are part of science fiction, not science. People would understand this if they weren't indoctrinated with the ridiculous SM being taught today as a result of superstitious magical foo held over from ancient times.

"We humans are so powerful we can kill the aliens with a mere thought."

Wow! And how do we "know" this?

"There is evidence in ancient rock carvings, ancient writings and aboriginal legends that the aliens made contact with humans in the distant past. … That we show different skin colors and appearances also suggests that the experimentation either was done by more than one visiting alien race…"

Evidence and experiments. Yep, that's good enough for me!

"But there is something even more important. It seems that we exist for a while in solid bodies that living creatures in sub-space do not have. And while we exist in them, our bodies give us the ability to touch, taste, smell, hear and fully enjoy the things around us."

Exist is still not defined scientifically. It's no wonder we hear this sort of thing all the time.

Psychic powers, parallel universes, alien and faster than light spacecraft, light beings that can inhabit our bodies, these are all things that persons believe based on evidence, true, false, observation and experimentation.

Once one understands the Rational Scientific Method, they will never fall for this sort of thing again. Why? Because what is possible or not possible is not dependent upon observation, experimentation, or belief of man!

If you are thinking well, yeah, of course those things are silly. That is only because you don't have the evidence before you that YOU need, unlike these others. Obviously, they have all the evidence that THEY need. Evidence, facts, true, false, and belief are all observation based...just like experimentation.

Science has its Many Worlds interpretation, where science fiction has its parallel universes.

Science has its mass-less particles, where science fiction has its ghosts.

Science has light particles that arrive before they leave, where science fiction has its time travel.

What's the difference between science and science fiction? It's hard to tell, if you don't understand the Rational Scientific Method.

Chapter Seven – Peer Review

Is it Scientific? What is peer review (in science)?

Peer review is supposed to be an evaluation of individual hypotheses or theories by other persons skilled in the same applicable area of science. The reviewer looks mostly for irrational, ambiguous or contradictory use of key terms and for inconsistencies in use of those terms throughout the presentation, then offers advice on how to improve it.

If the reviewer finds that all elements of a hypothesis are there, and that the theory explains the hypothesis without contradiction, then the conclusion would be, "The theory is a rational explanation of x," and the paper should be approved for publication. The reviewer makes an impartial evaluation of the article presented for peer review and offers it up for publication in a journal where it can be reviewed by interested parties in the appropriate field(s).

That's how it should work. That is not how it does work, unfortunately. Scholarly papers are routinely returned or filed in the circle file and ignored. Often the author's work is denied publication, not on the basis of an improperly prepared paper or "invalid" theory, but because the ideas of the presentation go against the status quo.

Persons with new ideas that go against current accepted theories are very often marginalized or labeled a crank. There are even websites dedicated to this purpose, such as crank.net.

"The free, unhampered exchange of ideas and scientific conclusions is necessary for the sound development of science, as it is in all spheres of cultural life." - Albert Einstein

Here are the comments from the Comment Section of a blog I started:

An Alternative To Waves And Wave Packets: vixra.org

"The mathematicians believe that we have already settled complex topics such as the splitting of the atom, the duality of light, the definition of a sphere, the physical meaning of positive and negative, the physical nature of a field and of space. The peer reviewers no longer ask authors questions about these topics. They ask new authors what their credentials are. Authority acts like a product warranty, a way to ensure that the editor doesn't waste valuable time reviewing crackpot theories. The peer reviewer wants to know if you went to the same seminary he did and whether you learned the same nonsense he learned. He wants to know if you are going to tell his readers what they have already been conditioned to believe. In other words, the contemporary mathematician has none of the virtues of a scientist.

"The peer review mechanism:

"The purpose of the peer review process should be to certify that a theory follows the scientific method (i.e., whether the paper is written in a sequence of rational steps). A reviewer should objectively check whether a theory follows from the premises.

"You should also be able to know your accusers. Thus, if a reviewer rejects your manuscript arguing that it is unscientific, you have a basis to sue for plagiarism if you later read your article under that juror's name in another magazine. Instead, the contemporary peer review process is conveniently kept secret. The logic behind this is flawed. There is no reason to keep the reviewers in anonymity other than for this dark inquisition board to judge the merits of your case (i.e., determine whether your paper conforms to the established religion). Ask yourself: if the peer review process really works, why are you still reading super-natural and irrational poppycock such as time travel, black holes, and annihilation in allegedly scientific journals and magazines such as Nature, Science, and Scientific American? You should be reading these fairy tales in fiction mags! Or perhaps we should rename Nature as Fantasy and Science as Science Fiction!" - Bill Gaede

Posted by Dr Strangelove : "Peer review of Giordano Bruno lead to him being burned at the stake in 1600 ad. Similar peer review of Galileo in 1633 led to him being forced to his knees to renounce all belief in Copernican theories, and was thereafter sentenced to imprisonment for the remainder of his days."Mainstream science, fortunately, no longer burns people at the stake but I think many of them would like to.

"Before even getting a paper peer reviewed, one must get it past the editors. I have had papers rejected not because of errors, but because of subject matter 'challenges accepted theories'.

"One problem is the interplay of business with academia. Journals are for the most part "for profit" businesses. Reviewers are for the most part customers of the journal. The journals get their customers to work for them for uncompensated. What a deal for the journals. Does anyone know of any publicly traded journals? If so, I am buying stock. The reviewers are usually University professors and get brownie points from their respective Universities. The higher a professor is regarded by his/her academic peers the more the University can charge for tuition. The reviewer and the journals then have no incentive to tip the apple cart, so to speak.

"Another problem is the power the academic world grants to certain individuals. Consider the Copenhagen Doctrine. A group of physicist tired of hearing new and different explanations for quantum observations got together and decided to accept a set of explanations thereby closing the books on any further theoretical progress. Imagine if, in the early days of personal computer development, Bill Gates, Steve Jobs, and the now litter known developers of the other early computers and software got together and decided by 'Doctrine' just how a PC should be built and programmed. Life would certainly be easier. They could fire the engineers and just concentrate on building and selling PCs. They would all agree on the doctrine and then rush back to their labs, hire more engineers and try to develop a machine better then the doctrine dictated while hoping their competitors were dumb enough to stick to doctrine. In physics there is nothing to sell and no clear and obvious financial gain from challenging doctrine. Rather there is potential financial damage one could be subject to by challenging doctrine.

"When I submit a paper, I would like to be reviewed by a peer of mine. If some open minded imaginative intelligent person submits a paper do they get a review by a 'peer'? No! They get reviewed by some closed minded, highly biased, fat lazy college professor.

"Despite it all, the truth will find its way into the mainstream."

John Baez and the Crackpot Index:

His paper scores pretty high on the crackpot list!

One arrogant fellow of the establishment who pokes fun at dissidents is John Baez. Some time in the 90's this individual created his Crackpot Index, a site that is mirrored in many conspicuous places and is read by tens of thousands worldwide. On his page, Baez suggests a grading system to determine who should be called a crank or a crackpot. A crank is a person with eccentric ideas which supposedly have already been debunked, but who against all odds, insists on them."

Chapter Eight - Proof Is For Alcohol

The earth was conceived to be a sphere hundreds of years before the telescope and 1700 years before Magellan's circumnavigation.

The Greeks conceived of a spherical earth in the 6th century B.C. but it took three centuries before (Hellenistic) astronomy's spherical earth was adopted.

In 1519-1522 Magellan circumnavigated the earth "proving" the earth was round, but it was 1700 years earlier that Eratosthenes used trigonometry to "predict" it.

So much for prediction, so much for proof, and so much for observation.

Spherical earth could just as easily have been conceived before the flat earth, and probably was. The prevailing opinions ruled the day, and man's need to observe, to know, and to prove took precedence over rational thought.

There are some things that can not be observed by man's limited senses, does that mean they are not there? For example, we can only use critical thought and our rational mind to conceive of the invisible things which mediate light and gravity.

We can only conclude that something is possible or not possible based upon our unlimited intellectual ability. Critical thought and the rational mind are the best tools that we have to explain and understand anything we can conceive of.

Chapter Nine - Rational Physics

- Objects have shape
- To exist is to have shape and location
- Space can not become matter and matter can not become space
- Matter and motion are eternal
- All phenomena are the result of surface to surface contact between objects

As discussed in Chapter One - Rational Scientific Method, all words fall into one of two categories.

There are objects and there are concepts. It is very important in any conversation to understand the difference. In science it is crucial that all terms, which make or break an argument or presentation, are defined scientifically in the hypothesis.

Put in terms of physics, there are physical objects and phenomena. Physical objects, that is, objects which exist, have shape and location. Physics is the study of physical objects; something somewhere, objects with location. A circle is an object because it has shape. It can be illustrated. It does not have location with respect to all other existing objects. Draw a circle on a piece of paper with a pen. What exists is the ink and the paper.

The circle is an abstract concept geometers invented. The shape is defined by that which allows us to conceptualize what is inside and outside of a border; space. The one criteria common to all objects is shape. This is an either/or situation. Objects have shape, space does not. Therefore, space is not an object!

Cut a circle out of a piece of paper and see that we still have a "shape," but it is now a ring of paper. The ring has the mutually orthogonal directions of length width and height; the three dimensions of reality. The ring exists because it has location with respect to all other existing objects.

When the paper ring was cut out of the piece of paper there was a hole, or space 'left' in the piece of paper where it was removed. Did we destroy paper? Did we create space from paper? Put the ring of paper back. Did we create paper from space? No, of course not, we simply changed the shape. Matter can not be created or destroyed. It is eternal! By matter I mean the sum of all objects being comprised of atoms.

Object: that which has shape

Space: that which lacks shape

Without the concept "space" we would not be able to conceptualize any existing objects because there would be no border or separation between them. We can take our ruler and measure the hole in the paper. We have not measured space; we have measured distance from one place on the paper to another place on the paper.

Also, if there was no space, or separation between objects, motion would be impossible. If we accept the definition for object, we see that all objects are finite. If there was no limit, or border to an object, no other objects would be possible, and therefore there could be no motion. What is motion? It is two or more locations of an object.

It is apparent that there is matter and motion. This is the default position one must take. If anyone says there was a beginning to matter (creation) or motion (first cause) then the onus is on them to show how this can be. Naturally, we understand that matter and motion are eternal. For how can nothing, that which lacks shape, or space, become some "thing." In other words, how can zero dimensions instantly become the three dimensions of reality? As we understand from our example above about paper, it is impossible for space to become matter and matter to become space. Also, how can there be a beginning to motion? If it is necessary for a first cause of movement, what moved the first mover? We go on forever recursively and never get to a first cause!

Lastly, we understand that all phenomena are the interactions between objects. We drew a circle on paper and we cut out the circle with a pair of scissors. The objects were pen, paper, and scissors. The phenomena were draw and cut. There could be no draw without paper and pen. There could be no cut without paper and scissors. There can be no phenomena without two or more objects interacting. The interaction involves contact between surfaces. Can you name an exception?

Chapter Ten - Experimenter's Regress

Reading through the Wiki article on Experimenter's Regress we find the following:

Certainty/knowledge "The outcome of a phenomenon that is studied for the first time is always uncertain and judgment in these situations, about what matters, requires considerable experience, tacit and practical knowledge."

Right/wrong "The scientist, in other words, has to get the right answers in order to know that the experiment is working, or to know that the experiment is working, to get the right answer."

Consensus, Quackademia and the castle guards "In new fields of research where no paradigm has yet evolved and where no consensus exists as what counts as proper research, experimenter's regress is a problem that often occurs. Also in situations where there is much controversy over a discovery or claim due to opposing interests, dissenters will often question experimental evidence that founds a theory."

Subjectivity "Because for Collins, all scientific knowledge is socially constructed, there are no purely cognitive reasons or objective criteria that determine whether a claim is valid or not."

Persuasion/belief "Acceptance of claims boils down to persuasion of other people in the community. Experimenter's regress can always become a problem in a world where "the natural world in no way constrains what is believed to be"."

In summary: "Sextus Empiricus' argument that "if we shall judge the intellects by the senses, and the senses by the intellect, this involves circular reasoning inasmuch as it is required that the intellects should be judged first in order that the intellects may be tested [hence] we possess no means by which to judge objects" -

Add certainty, knowledge, right, wrong, consensus, subjectivity, persuasion, and belief...mix well; add authority and you have the reason why science fails majorly.

Nothing is certain, there is only possible and not possible. Right and wrong are a matter of opinion. Consensus is merely the opinion of many. Subjectivity relies on the limited, often faulty human senses. Persuasion, belief, and authority are part of religion and philosophy. These have no place in the business of scientific inquiry.

Couple confirmation/hindsight bias with inductive/deductive and circular reasoning and you wind up with abstract concepts like Relativity, Quantum Mechanics and String Theory instead of objects explaining phenomena.

Experimenter's regress is a given considering the current model of induction, deduction, and circular reasoning.

The only way to break out of the loop is to: "Kill the observer," that is, eliminate subjectivity.

How to Eliminate Subjectivity

The Rational Scientific Method is a process of eliminating subjectivity.

We observe something, and that is what leads us to the scientific method. We use the hypothesis to conceptualize the objects.

We define so that everyone understands what we are talking about. In this way we may enlist the intellects of others to help us eliminate, or fine tune our assumptions.

If the hypothesis fails, we erase the whiteboard and start on something else. If we have a viable hypothesis, we can move on to the theory which explains the phenomena.

The conclusion of possible or not possible is no concern of science. If the theory is possible, the theorist can move on to the next hypothesis, and the engineers and technologists can take over. They may decide to apply the theory towards building something useful.

This is when empiricism comes into play. Engineers, technologists and tinkerers can test, experiment, and use trial and error to their hearts content.

If not possible, there is no big loss of money or time. Since no experimentation has been done up to this point, little or no money has been spent on materials, lab time, research, accounting or personnel. This eliminates the emotional attachment to an outcome due to financial or other obligations.

The reason science and technology are separate departments is because of this: Science explains, technology builds. However, they are not interdependent as claimed. They are independent. There are literally thousands of scholarly articles spelling out the differences, and the rivalry between departments.

One would think that if science and technology are interdependent yet separate that there would be a friendly relationship and a clear division of labor between the two. Yet this is not the case. Scientists, engineers, social scientists, philosophers, historians, policy makers, and the public all have self interests. Because of this, government and other agencies, like NASA, have Science and Technology departments just to help solve related issues. They also coordinate resources and development across the disciplinary boundaries.

Science is conceptual, technology is empirical, and never the twain shall meet! The theorist needs to stay out of the laboratory.

Evidence, proof, and facts are not scientific. These are opinions. Experiments are not part of the Rational Scientific Method of inquiry because they are dependent upon the limited sensory system of man and are therefore considered opinion. We reserve our opinions for after the hypothesis and theory is presented. Our opinion is called the conclusion (which is ours alone).

Technology is not a practical application of science. A theory is an explanation, while machines, devices, and widgets are built mostly by trial and error (and very many inventions are pure happenstance).

We may be told that the scientific method is designed to help us harness luck and make discoveries. Louis Pasteur said, "Chance favors the prepared mind."

We saw in a previous chapter many examples of what we may have considered as science, but were accidental discoveries. As I said before, while those are all good examples of accidents, luck and happenstance meeting up with open minded, observant men and women, it points to something very important. It points to the difference between science and technology, conceptualizing and observing.

The purpose of science is to explain and that depends on conceptualization. The usefulness of technology is in designing and building. This is accomplished by experimentation, testing, and trial and error which depend on observation.

Clearly, science and technology are two different things altogether. The problem with modern scientific method is that they confuse the two.

To find out more about why true/false, right/wrong, proof, belief, evidence, and authority are NOT part of the Rational Scientific Method of inquiry, read the entire series of Rational Science books.

Chapter Eleven - Knowledge and Prediction

"The meaning of knowledge is known to all, and I think it means acquaintance with a fact, science or technique for example." - Andy

- Fact: opinion from authority
- Authority: Self appointed expert on a particular subject

"I see what you are insinuating. Thanks." - Andy

Typically we see something like this from Webster's on-line dictionary:

"knowl·edge - facts, information, and skills acquired by a person through experience or education; the theoretical or practical understanding of a subject."

If one can rationally explain it, then they understand it. However, much of what self-proclaimed authorities and scientists claim as knowledge is nonsense. How can anyone possibly understand the irrational (impossible)?

Mainstream physicists use "predict" and "know" irrationally.

When they say "equations predict," such as, F=ma predicts how fast an apple falls from the tree, they are actually describing consummated events.

When they say they know the apple will fall from the tree at 9.8 ms ^2 and then hit the ground, they mean they 'predict' based upon past experience. If a raven swoops down and grabs the apple before it hits the earth, so much for their knowledge, and so much for their prediction.

Our "knowledge" is based in large part on observation, and observation is subjective, being built from limited human senses.

Because of this and because of "an ongoing collision between belief and reality" science necessarily must remove the observer from the scientific method. Knowledge, experience, and belief are opinions, and have no place in science or physics which studies what is physically real.

Knowledge has nothing to do with reality. Knowledge is a hallmark of religion. Knowledge just means that you have made up your mind about something or other. There is only one way for you to show that you know something: you run an experiment and prove your knowledge to yourself.

- know/knowledge: The ability to predict the result of an experiment without error. Knowledge only has to do with the future.
- explain/explanation: To state the causes of a phenomenon (consummated event). Explanation only has to do with the past.

Science is the study of existence/reality using the Scientific Method. There is no provision for knowledge in the scientific method.

The scientific method only deals with consummated events which are rationally explained. Astrology deals with future events which are predicted and allegedly known. In science we strive to understand. We do not confuse understanding with knowledge.

"The greatest enemy of knowledge is not ignorance, it is the illusion of knowledge." -- Stephen Hawking

See also, Rational Science Vol. IV, Chapter Twenty One – Relativity's Failed Predictions

Chapter Twelve - Word Magic V1.1

Tired of being pushed into the corner with scientifically accurate words and phrases? Well, no more, astound, befuddle, and bamboozle your opponents with a working knowledge of the top misleading words in the human lexicon. With words like energy, field, wave, and life you'll never be pinned down again!

With WordMagic V1.1, you can say anything, and it is guaranteed to be true, correct and factual!

You can be the life of the party! You can be unequaled in irrelevancy, fabulously fallacious, and deliciously discrepant. Learn how to impress your family, friends and neighbors with counterfactual vernacular and meaningless jargon.

Whether at the seminar or symposium, conference or confabulation, stun your co-workers, colleagues, and collaborators, with inconcise, indeterminate and unspecific Magical words of misrepresentation and miscommunication.

With WordMagic V1.1 you'll be able to define ambiguously, circularly, synonymously, and inconsistently thousands of authoritative, popular words from our tested and true Word Magic Lexicon.

But Wait! There's more! Order now, and we'll include, for no additional charge, our very specially narrated CD version. Listen to the smooth, sonorous voices of folks like, Stinking Hawking and Dick Dawk as they teach you how to dodge and weave through any presentation, debate, conversation or convocation.

And we're still not done! For the next 100 persons who order WordMagic V1.1, we'll include, free of charge, the additional Musical CD, "WordPlay". Relax during an office power nap, or send yourself off to Slumberland listening to these favorite hits: Double Talk, Ambiguity, Dubiety, Incertitude, Tergiversation,

Equivication, Polysemy, Double-entendre and Enigma.

WordMagic It's not just for scientists anymore!

Chapter Thirteen - Nature of Scientific Inquiry

Why We Need a Rational Scientific Method

A paper at academia.org has come to my attention. I find it very interesting and revealing of how academia approaches the "nature of scientific inquiry." There is a movement towards science education reform by way of educating the educators away from "naive views" of objectivity. Research has been underway to study the conceptions teachers have about scientific inquiry. The author of the paper notes:

"Some naïve ideas harboured by the teachers are that if imagination and creativity were to be used in science, then science would lose its worth of being a body of facts."

The author of the paper, Washington Dudu, considers this a "naive realist argument" of the...

Nature of Scientific Inquiry

The paper is not about the Scientific Method of Inquiry, per se, it is about educators conceptions of the nature of scientific inquiry (NOSI).

Dudu tells us:

"Conceptions of the nature of scientific inquiry are an individual's ideas, beliefs, understanding and assumptions about the scientific process; what scientists do; and how scientific knowledge is developed and validated (Vhurumuku & Mokeleche, 2009)."

The study elicited five teacher's conceptions of the nature of scientific inquiry utilizing the three tenets (assumptions) of NOSI:

(1) scientists use a variety of methods to conduct scientific investigations
(2) scientific knowledge is socially and culturally embedded; and
(3) scientific knowledge is partly the product of human creativity and imagination

This study endeavors to discover what, among the science education community, are the beliefs, opinions and "doctrines" about science and the scientific process.

The study's questionnaire was also based on (but not discussed in the paper) three other tenets:

4) difference between laws and theories

5) accurate record keeping, peer review, and replicability in science; and

6) theory ladeness of observations.

The first three tenets spell out perfectly well the problems inherent with the mainstream scientific method of inquiry enumerated in the last three. Nature does not operate based on any of man's laws. Hopefully, one can tell the difference between the hypothesis and theory, thus eliminating any need to discuss the difference between "laws" and theory.

One presents their assumptions, defines their Key Terms, and illustrates or describes the objects in their hypothesis. The phenomenon is then explained in the theory. Science is done! Each individual forms their own conclusion as to whether or not the explanation is possible.

Record keeping, peer review and replicability are based on observation, and therefore are inherently flawed. Observation is based on the limited sensory system of man and is therefore subjective. The scientific method should eliminate as much subjectivity as possible if its goal is to be objective!

Theory ladeness of observation is the idea that the theory is affected by presuppositions of the individual researcher, and takes on two forms:

a) The theoretical presuppositions determine the meaning of the terms; and

b) The perceptions are influenced by the theories held by the theorist

Empiricism naturally leads to opposing views such as sense data theory, physicalism and foundationalism. It gave rise to the inductivists, hypothetico-deductivists, and Popperian falsificationists. The modern mainstream scientific method of inquiry now utilizes inductive reasoning, deductive reasoning, and circular reasoning as detailed in the chapter, 'Scientific Method? For Dummies!' Rational Science Vol. 2

How do we avoid all these 'isms' and empiricist traps? By removing the observer! I suggest the Rational Scientific Method.

For now let's continue with the study. Under the heading 'Scientific knowledge is socially and culturally embedded', Washington Dudu tells us that four of the five teachers questioned in the study "held contemporary and informed views that science is part of social and cultural traditions."

When one of the participants responded with this:

"Science is based on facts. Most of the time, science proves social and cultural beliefs held by individuals wrong. In the olden times, people believed that the sun moved around the earth, but Science proved that it was the earth that moved around the sun [...]. "

Dudu says:

"This view is categorized here as naïve. Though [the] argument is reminiscent of the argument peddled with much emotion by Copernicus and the Christian religion, according to Inokoba, Adebowale and Perepreghabofa (2010), science as an endeavour and phenomenon is not conceived and operated in a cultural and environmental vacuum. It is a social phenomenon greatly influenced by the prevailing cultural traits and worldview of a people such as their social values, priorities, ideas, skills, ethics, perception of social reality, and belief systems."

Under the heading, Scientific knowledge is partly the product of human creativity and imagination, we find four of the five teachers "believe that science knowledge and truth are not fixed by nature but are also creations of the mind" and, we are told, "These views are categorised here as informed."

The fifth teacher disagreed that science was a "product of creativity and imagination" saying those things conflict with objectivity. The author of the study thinks that this is a naïve view of NOSI, and describes it as a realist view.

Dudu says this:

"[The participant's meaning of the term "objective" appears to be the same in this instance as that for the term real. To [The participant], if an idea cannot be proven, then it is not scientific and not real. [The participant] has to be reminded that scientists do not solely rely on logic and rationality. In fact, both creativity and imagination are a major source of inspiration and innovation in science. It permeates the ways scientists design their investigations, how they choose the appropriate tools and models to gather data and how they analyse and interpret results."

It would be very interesting to see how Dudu defines the terms real, objective, and proof. Of course, science does not prove anything, and doesn't claim to, so it is important to know if the participant actually thought this, instead of "appeared to think" this as stated by Washington Dudu. While creativity and imagination can be great tools for technology with its empirical approach, it has no place in rational science. Are we to take a show of hands on whose tools, whose models, and whose data we are to analyze? And whose interpretation are we to rely upon? Are we to accept the consensus of an authoritative panel of experts? Nature recognizes no such panel or authority.

Finally, we are told:

"Intricately linked to the subjectivity of science is the use of imagination and creativity in scientific investigations. Some naïve ideas harboured by the teachers are that if imagination and creativity were to be used in science, then science would lose its worth of being a body of facts. This is a naïve and realist argument put forth by some of the teachers."

Dudu states, "As a human endeavour, science is influenced by the society and culture in which it is practised" and (as summarized by Liang et al) "cultural values and expectations determine what and how science is conducted, interpreted and accepted." While it is naïve to think that science is not influenced given the mainstream method of scientific inquiry as presented in tenets four, five and six listed above, it is not naïve to expect that science should be objective and therefore free from imagination, creativity and any particular scientist's cultural values or worldviews.

With a Rational Scientific Method of inquiry which removes the observer, cultural traits, social values, and belief systems will have no influence on the process.

The paper, Exploring South African high school teachers conceptions of the nature of scientific inquiry a case study can be found on Academia.edu.

Chapter Fourteen - Karl Popper

The Definition of Crazy

"No rational argument will have a rational effect on a man who does not want to adopt a rational attitude." — Karl Popper

"No hypothesis will have a rational theory if one does not adopt a rational scientific method." – Monk E. Mind

While this is in no way intended to be a thorough examination of the man or his Critical Rationalism (nor did I exhaust the resources available on the subject), a fair analysis can be made if we are to accept Karl's own words as representative of his position. I quite honestly never came to a clear understanding of his Critical Rationalism, or other philosopher's critique of same, and anyways prefer to keep the main thing the main thing, focusing primarily on falsifiability.

Quotes gleamed from the following references:

Stanford Encyclopedia of Philosophy

Karl Popper and Critical Rationalism - wikipedia

Nicholas Dykes; A Critique of Karl Popper's Critical Rationalism - Reason Papers

Modern mainstream science has incorporated Popper's falsifiability into its scientific method along with induction, deduction, and circular reasoning. It is not difficult to notice that philosophy had, and still has, a great influence on the sciences.

Karl (May I call you that? Thank You!) Karl was critical of induction. Yet, as one of his critics, Nicholas Dykes, fairly points out:

"Collecting disconfirmations and arguing negatively scarcely differs from collecting confirmations and arguing positively. Both are inductive procedures and, as such, have been disallowed in advance by Popper's rejection of induction. The bottom line which CR [Ed. Critical Rationalism] must confront, however, is that one cannot falsify a scientific theory without inference from observed instances. However much Popper may have rejected induction, his own method was in fact dependent upon it."

And

"If our senses are automatically suspect, as Popper maintained, negative or falsifying instances deserve no more credibility than positive or confirming ones."

A critic, Anthony O'Hear, said, "There can, in fact, be no falsification without a background of accepted truth."

Dykes and other philosophers recognize the true/false dichotomy, yet miss the most important dichotomy in science: object/ concept. That's what we can expect when we have priests and philosophers running the science department.

Karl was a mixed bag of incrogruencies. He disagreed with Niels Bhor's instrumentalism and Copenhagen's interpretation of QM (Quantum Magic) yet supported Einstein's realism. Instrumentalism sees a theory as useful in as much as it explains and predicts phenomena; realism says that reality is what it is regardless of what we think it is.

Popper proposed a physical experiment similar to Einstein's thought experiment (EPR) because he saw non-locality as contrary to common sense and wanted to falsify action at a distance. However, his experiment called for counting particles and particles are the bane to both QM and Relativity alike.

Popper stated that "There is no such thing as an unprejudiced observation" and that "our scientific theories must always remain hypotheses." He also said this:

"The quest for certainty is mistaken .... though we may seek for truth ... we can never be quite certain that we have found it"

"No particular theory may ever be regarded as absolutely certain .... No scientific theory is sacrosanct ..."

"Precision and certainty are false ideals. They are impossible to attain and therefore dangerously misleading ..."

If reality is what it is in spite of what we think it is, and "We never know what we are talking about then why not remove the observer from the scientific method of inquiry? Instead, Popper doubled down on testing and experimentation because, he said, we could never know with certainty if our theories are correct, we can only apply probabilities to their correctness.

"Good tests kill flawed theories; we remain alive to guess again." - Karl Popper

MATHEMAGICS and TRUTH

"A statement is true if and only if it corresponds to the facts."

"Our aim as scientists is objective truth; more truth, more interesting truth, more intelligible truth. We cannot reasonably aim at certainty. Once we realize that human knowledge is fallible, we realize also that we can never be completely certain that we have not made a mistake." — Karl Popper

Not only was Popper a philosopher, he was a mathemagician. Karl had a formula for determining the falseness and 'truthiness' of a theory:

"Informative content, which is in inverse proportion to probability, is in direct proportion to testability."

$$Vs(a) = Ct_T(a) - Ct_F(a),$$

where $Vs(a)$ represents the verisimilitude of a, $Ct_T(a)$ is a measure of the truth-content of a, and $Ct_F(a)$ is a measure of its falsity-content.

Although Karl's theory of verisimilitude was seen as deficient, and not widely accepted at the time, it was considered central to his "philosophy of science," and today is reflected in the teaching of students that their "conclusions will always be tentative ones."

And a formula for Predictive ability:

[C.P. + E.S.]=U.P.

He saw the advance of scientific knowledge (wait I thought we could never "know' anything?) as an evolutionary process.

He had a formula for that too:

PS1 > TT1 > EE1 > PS2

It would seem that Popper's demarcation theory (how we separate science from non-science with falsifiability) would invalidate mathematics, physics, and logic as scientific.

Popper's philosophy of mathematics solved the question of how statements of math like 2+2=4 could never be shown false. By his own account, it is not scientific if it can not, in theory, be proven false. How can we learn about reality from it, then?

Simple, says he, the statement "2 apples + 2 apples = 4 apples" has two applications. It is logically true (tautology- axiomatically true) but in reality it may also be falsified.

Karl Popper has the distinction of bringing falsifiability into the scientific method of Popperlar Science.

So, what of this falsifiability? What ever did he mean by that?

"Nothing in the empirical sciences can ever be proven, but it is falsifiable, that is, it can and should be scrutinized by decisive experiments. The term "falsifiable" does not mean something is made false, but rather, if it is false, it can be shown by observation or experiment."

"In so far as a scientific statement speaks about reality, it must be falsifiable; and in so far as it is not falsifiable, it does not speak about reality." - Karl Popper

According to Stanford's Encyclopedia of Philosophy:

"For Popper, a theory is scientific only if it is refutable by a conceivable event. Every genuine test of a scientific theory, then, is logically an attempt to refute or to falsify it, and one genuine counter-instance falsifies the whole theory."

Sounds like my definition for crazy- doing the same thing over and over again expecting a different result.

What about definitions? Glad you asked! Karl Popper didn't like definitions. Apparently he loved word magic!:

"One should never get involved in verbal questions or questions of meaning, and never get interested in words. If challenged by the question of whether a word one uses really means this or perhaps that, then one should say: "I don't know and I'm not interested in meanings......one should never quarrel about words, and never get involved in questions of terminology. One should always keep away from discussing concepts."

"Definitions do not play any very important part in science .... Our 'scientific knowledge' ... remains entirely unaffected if we eliminate all definitions."

"Definitions never give any factual knowledge about 'nature' or about the 'nature of things.'"

"Definitions .... are never really needed, and rarely of any use." - Karl Popper

Poppercock!

"Anyone who refuses to define uses word magic!" – Monk E. Mind

"Not to have one meaning is to have no meaning, and if words have no meaning, our reasoning with one another, and indeed

with ourselves, has been annihilated. - Aristotle
John Herman Randall, Jr.,

"Science must begin with myths, and with the criticism of myths." - Karl Popper

God and Exist

"I don't know whether God exists or not. ... Some forms of atheism are arrogant and ignorant and should be rejected, but agnosticism—to admit that we don't know and to search—is all right. ... When I look at what I call the gift of life, I feel a gratitude which is in tune with some religious ideas of God. However, the moment I even speak of it, I am embarrassed that I may do something wrong to God in talking about God.

"Natural laws .... do not assert that something exists or is the case; they deny it.

"If we call the world of ... physical objects ... the first world, and the world of subjective experiences ... the second world, we may call the world of statements in themselves the third world." - Karl Popper

Well, if Karl would have only defined the term EXIST, he would perhaps have had a different take on this. Critical to the scientific method at the hypothesis stage is defining one's Key Terms.

Exist: object with location; something somewhere; physically present

I propose to you that science is never about true or false, these are questions to ask your priest or philosopher, not your scientist!

Science is about explaining, and the conclusion is yours to make: Possible or NOT possible

Please take another look at the first chapter in this book, "Rational Scientific Method."

Chapter Fifteen - Words Mean Things

Definition and Context are Important!

"It depends on what the meaning of the word 'is' is." – President Clinton

At the 'higher' level of government, politicians understand full well the importance of defining one's Key Terms. Take for example the term 'coup.' One would think that it would be a simple matter of determining whether or not what is going on in Egypt (2013) is related to a coup, if the word is defined.

The most common meaning of coup being the following:

From oxford dictionaries.com

Coup: (also coup d'état) a sudden, violent, and illegal seizure of power from a government: he was overthrown in an army coup

It is obvious to me that U.S. officials ARE familiar with the definition, and also why they would wish to have an ambiguous definitions for THEIR key terms.

With accusations that Israel is behind the military coup in Egypt, and the amount of money promised (1 billion) at stake, it is no wonder US politicians want to consider re-defining the word 'coup.'

"If Morsy's removal were to be called a coup, under U.S. law, more than $1 billion in military aid to Egypt would have to be slashed."

"Senior U.S. officials say the administration is examining three potential options – calling events in Egypt a coup and cutting off aid; calling it a coup and issuing a national security waiver; or not determining it a coup, recognizing that the military has taken steps to move the country toward a civilian transitional government and move toward elections."

"So our decisions with regards to the events that have happened recently in Egypt will be - and how we label them and analyze them will be made with our policy objectives in mind, in accordance with the law and in accordance with any consultation with Congress," he (Carney) said."

From a CNN blog July 8th, 2013 entitled U.S. Avoids Calling Egypts Uprising a Coup

There are also political motivating factors in defining, or shall I say- not defining, or defining ambiguously, certain terms in science.

If physicists were forced to define in unambiguous terms their magic words energy, field, space, and time, their whole house of cards would collapse, and money for projects like CERN and Fermilab would dry up over night. Careers would come to a standstill, funds would dry up, and labs would be shut down around the world.

Besides the obvious political motivation(s) behind defining terms, or using "labels," as our friend Mr. Carney called it, there are other (apparently) less obvious reasons for our particular use of words. One may be predisposed to thinking about certain things in certain ways depending on their language and their culture. Who would of thunk it?

From Psychology Today: How Culture Shapes Thought by Lawrence T. White, Ph.D., and Steven B. Jackson

"The evidence is clear: To a surprising degree, language and culture influence how we think about time."

Wow! Startling! Who would have guessed? A number of 'scientific' studies by psychologists, linguists, and anthropologists have come to this startling conclusion based upon a number of recent experiments.

Let's look briefly at some of the evidence and conclusions about how and why persons use the words they do for time and space. We'll draw from two seminal articles on the subject:

Does Language Shape Thought?: Mandarin and English Speakers' Conceptions of Time by Lera Boroditsky

And

The Thaayorre Think of Time Like They Talk of Space by Alice Gaby

Boroditsky notes that English speaking, and Mandarin speaking individuals talk about time differently. Whereas English speaking folks refer to time as 'moving' from left to right, Mandarin refers to time as 'moving' from up to down. In other words, if asked to place January and February in order, English speaking individuals place January to the left of February and Mandarin speaking individuals place January above February.

Of course, everyone notices that English reads and writes from left to right and Mandarin from top to bottom. The experiments looked at both monolinguals and bilinguals. It was concluded:

"(1) language is a powerful tool in shaping thought about abstract domains and (2) one's native language plays an important role in shaping habitual thought (e.g., how one tends to think about time) but does not entirely determine one's thinking in the strong Whorfian sense.

Key Words: Whorf; time; language; metaphor; Mandarin."

Note: the Key words listed were not defined, apparently just used to draw Google hits.

That article asks questions, such as; Do different people who speak different languages look at the world differently? Do people who speak more than one language think differently when speaking a different language? Is thought determined entirely by language?

Boroditsky does acknowledge the inherent difficulties related to the experiments. Of course, I consider experimentation extra scientific, so will not focus primarily on the experiments themselves, and data analysis, etc. My intention is to merely mention the particular ways in which it has been observed that different language speaking persons talk about space and time and the resulting conclusions arrived at by the experimenters.

According to the studies, experiences about the phenomena of time inform individuals that any particular event happens only once (sorry no Ground hog Day) are unidirectional, and this appears to be universal across differing languages and cultures.

"However, there are many aspects of our concept of time that are not observable in the world. For example, does time move horizontally or vertically? Does it move forward or back, left or right, up or down? Does it move past us, or do we move through it? All of these aspects are left unspecified in our experience with the world. They are, however, specified in our language—most often through spatial metaphors. Across languages people use spatial metaphors to talk about time. Whether they are looking forward to a brighter tomorrow, proposing theories ahead of their time, or falling behind schedule, they rely on terms from the domain of space to talk about time (Clark, 1973; Lehrer, 1990; Traugott, 1978). Those aspects of time that are not constrained by our physical experience with time are free to vary across languages and our conceptions of them may be shaped by the way we choose to talk about them. This article focuses on one such aspect of time and examines whether different ways of talking about time lead to different ways of thinking about it."

If the experimenters would have had time defined their Key Terms unambiguously they would understand how ridiculous those questions are. Those "aspects" are unspecified in our experience but are specified in our language precisely because time has not been defined scientifically, explained, or understood.

Different ways of talking about time naturally lead to different ways of thinking about it. Defining terms is paramount in understanding what people are talking about.

English speakers use 'front/back words' to refer to time such as before, after, behind, and ahead; mostly ordering events by using the same words as those used for describing horizontal spatial relations."

Mandarin speakers also use spatial relations to refer to time, such as front and back, but they also use vertical metaphors, such as up and down (used less often in English).

Boroditsky asks: "Does Metaphor Use Have Long-Term Implications for Processing?"

The conclusions based on the experiments answered her questions:

"As predicted, English speakers answered purely temporal questions faster after horizontal primes than after vertical primes.

"When answering questions phrased in purely temporal earlier/later terms, Mandarin speakers were faster after vertical primes than after horizontal primes. This pattern was predicted by the fact that in Mandarin vertical metaphors are often used to talk about time."

Did we really need to do experiments, along with their descriptive statistics and analysis, to arrive at that conclusion? How we think affects our speaking and how we speak affects our thinking. What is so mysterious about this?

How we habitually use words obviously affects the way we think. We didn't need experiments to understand this, we simply observe it regularly in our every day lives, and can conceive that this is the case with ALL words, not just space and time.

Our observations confirm our observations. Great! But how do we explain this? Words mean things! How we use words in conversational language is different than how we use our words in science, but it is ALWAYS important to understand the words we use.

Let's take a look at Gaby's article. We note that regardless of language or culture humans tend to speak about time in terms of space.

It's no wonder relativists believe that time combines with space because they think space is a substance. It is no wonder that relativists think of space as a substance because they reify concepts into objects, that is, they turn concepts (verbs) into the nouns of reality. Why? Because they were never taught to define terms unambiguously, and in non-circular, non-synonymous and non-contradictory ways. They were never taught NOT to reify, and they never learned to use their words consistently in their presentations. In short, they don't understand the difference between objects and concepts!

Let's continue. Gaby compares an Australian aboriginal group (both monolingual and bilingual) with English monolinguals to conclude the same thing that Boroditsky did; "The way people conceptualize time is shaped by a range of external influences, both linguistic and non-linguistic."

Other researchers determined the same thing as well. Hebrews order time from right to left, The Kuuk Thaayore from east to west, the Mandarin from up to down (or away to towards) and Tagbanwa from bottom to top. All this time/space mapping is according to a relative frame of reference (damn those relativists...they're everywhere!).

The interesting thing about the Kuuk Thaayorre is their mapping of time and space using an "absolute description of spatial relationships." They always map time from east to west no matter which way they are facing. If talking about a long time ago, they point to the east, for instance.

Their north and south frame of reference is anchored to their coast line.

"The terms -kaw "east" and -kuw "west" are defined by the sun's trajectory, while the terms roughly translated as "~north" and "~south" (-ungkarr and -iparr, respectively), more accurately align with an axis defined by the local coastline, forming an axis rotated almost 45° clockwise from that perpendicular to east-west."

They keep track of time using the sun and the moon, and also seasonal variations in flora and fauna. It is very interesting how they use sand paintings when telling stories. For instance, they may represent something to the east as in the past. They may erase something and draw over it for events that took place concurrently.

Another article of note by Boroditsky is entitled: How Language Shapes Thought, which has this to say:

"Language also appears to be involved in many more aspects of our mental lives than scientists had previously supposed. People rely on language even when doing simple things like distinguishing patches of color, counting dots on a screen or orienting in a small room..."

But this is the more salient point, I think:

"speakers of different languages also differ in how they describe events and, as a result, how well they can remember who did what. All events, even split-second accidents, are complicated and require us to construe and interpret what happened. Take, for example, former vice president Dick Cheney's quail-hunting accident, in which he accidentally shot Harry Whittington. One could say that "Cheney shot Whittington" (wherein Cheney is the direct cause), or "Whittington got shot by Cheney" (distancing Cheney from the outcome), or "Whittington got peppered pretty good" (leaving Cheney out altogether)."

Don't be fooled by the politician's or scientist's word magic! Always make them define their Key Terms.

Not only is it important to define one's key terms, it is also important to understand the context in which they are being used.

To find out more about why true/false, right/wrong, proof, belief, evidence, and authority are NOT part of the Rational Scientific Method of inquiry, read the entire series 'Rational Science.

Chapter Sixteen – Mathematics

The Language of Science? No!

I use some words differently than what one may normally encounter in communicating ideas relating to science. Please re-read the Rational Scientific Method to get a better understanding of what I am trying to convey.

When I use words in the scientific sense I will define them precisely in unambiguous, non-circular terms unlike conventional or contemporary science. I may refer to rational science as opposed to contemporary or mainstream science. I may refer to a physicist versus phizwhiz or mathemagician when making a distinction between rational physicists and mainstream theoretical physicists. These distinctions shall become readily apparent as one reads through the various chapters of this book. As far as rational science is concerned, the mathemagicians have been masquerading as physicists since Sir Isaac Newton got hit in the head with an apple.

Since that time phizwhizzes have made irrational claims, and their disciples nod in agreement like bobble heads without a clue as to the underlying flaw in their craft. They are not working with rational science. They have combined religion with science and reversed philosophy and physics. Whereas Rational physics is the

study of what is physically present, that is, what exists, mainstream physics is about the study of abstract concepts. We are told that the study of objects that exist is for the student of philosophy.

Mainstream or contemporary physics involves itself with such abstract concepts as Big Bang, Black Holes, spacetime, and Many Worlds Interpretation. The only way to understand whether the phizwhiz is doing physics or magic, is to understand the difference between concept and object, and then force them to define their key terms.

The Rock Star phizwhiz and his bobble heads use conversational language to describe abstract concepts and they do that inconsistently throughout their presentations. These individuals don't even realize that they have not communicated their ideas effectively. Instead, they will tell you that General Relativity, Quantum Mechanics, and String Theory are strange and difficult subjects - especially for the layperson. In rational science, the theorist understands the difference between concepts and objects, and can define words like 'exist' unambiguously, applying them consistently throughout the presentation of his or her hypothesis and theory.

Today the individuals of mainstream contemporary physics are mathematicians. The establishment uses Rock Stars such as Michio Kaku, Brian Green, Lawrence Krauss, and Stephen Hawking to popularize the language of theoretical mathematical physics. They sing their songs and the bobble heads nod in unison keeping a beat. As the right arm of mainstream public relations, they want you to believe that they are the authorities on reality. The aspiring young phizwhiz emulates these individuals and wants to grow up to be just like a Rock Star too. However, there are no authorities on reality. Please re-read the chapter on 'authority'.

Authority no longer comes from the ancient texts of religion, but from the whiteboards of mathemagicians in the form of equations.

It is very important to understand that mathematics is not the language of science or physics. Math is not even related to physics and has limited use in science. One should not interpret the physical world from abstract mathematical concepts and claim authority, anymore than one can read a Harry Potter novel and claim their interpretation is the authoritative one. While math is a language, it is not the language of science anymore than German, Latin, or English. Mathemagicians can no longer tell the difference between what is a real object of existence and an abstract mathematical construct. The bobble heads believe what they are told without questioning the rationality of it and revere the Rock Stars of science and Physics.

Mathematical equations are irrelevant to understanding reality. Equations can describe phenomena yet explain nothing. Newton's equation described an apple falling, yet could not explain why the apple did not fall up into the sky instead of down onto his head. He admitted that he had no hypothesis:

"I have not as yet been able to discover the reason for these properties of gravity from phenomena, and I do not feign hypotheses. For whatever is not deduced from the phenomena must be called a hypothesis; and hypotheses, whether metaphysical or physical, or based on occult qualities, or mechanical, have no place in experimental philosophy. In this philosophy particular propositions are inferred from the phenomena, and afterwards rendered general by induction." – Newton from Principia

General Relativity, Quantum Mechanics, and String Theory are the three legs on the stool used by mathemagicians for milking the government cow out of billions of dollars in research grants. They may prop up the phizwhiz, but are useless for explaining reality to the rest of us.

The Rock Stars of Physics feed their hungry cult following a large helping of irrational nonsense such as zero dimensional point particles and multiple universes, and they eagerly eat it up. When a rational person doesn't buy into it, they are told that they cannot use their common sense because the universe is not sensible. Then they parade a long line of 'authorities' and Nobel Prize winners in front of them and ask, "Who do you think you are to question authority?"

What has come of hundreds of years of mathematics? Not one mathemagician has been able to explain even the most fundamental aspect of reality such as light or magnetism.

We are expected to believe in, rather than understand, infinite worlds and irrational entities such as black holes or zero dimensional particles.

Contemporary science doesn't understand the scientific method! Not only do they confuse verbs with nouns and concepts with objects, they confuse hypotheses with theories.

The phizwhizzes and their bobble heads sadly believe that the inventions of technology prove their theories. If I've heard this once, I've heard it a thousand times, "GPS proves relativity," or "The transistor proves quantum mechanics," or some other such nonsense. These things and many other useful inventions are the product of human rationality, ingenuity, and the trial and error of technology. Advancements in technology are not a result of mathematical theories.

The first step of the scientific method is to illustrate or describe physical objects, define key terms, and make a statement of facts. This is the hypothesis. It requires objects, not concepts! Theory explains the hypothesis. Each person forms their own conclusion as to whether or not the theory presented is possible or not possible. Experiments and mathematical abstractions are extra-scientific. They are not a part of the scientific method.

The mathematical physicist takes abstract concepts and reifies them into the nouns of reality: they turn verbs into nouns. While this is ok in conversational language, it is not ok in science. They use undefined terms such as point, line, or plane, and then use these terms inconsistently in their presentation. They move abstract concepts and describe interactions between abstractions. Math can do anything that reality cannot! Math confuses thought space with thing space and mathemagicians cannot separate the two.

The job of understanding the nature of reality has been handed over from the priest to the mathemagician who offers no explanations whatsoever. They give us ridiculous ideas such as spacetime, parallel universes, and wavicles. They are the authors of time travel and warped space. Do we really want to continue to depend on these irrational, non-sensible mathemagicians to tell us what reality is? It's high time we leave the Dark Ages, and use the greatest tool we humans have - the intellect. The intellect is the ability and the willingness to conceive of concepts and apply them to reality.

Chapter Seventeen – Life

What Is It Anyways?

I asked four people in a science forum, "What is life?" and got four different answers:

"Everything is alive."

"To be alive is to have life...the vital ingredient to resist entropy."

"Things that are alive make copies of themselves."

"It depends on how you define life. A simple definition of life is that it can make copies and evolve."

The dictionary definitions were just as varied:

Usually something to do with what distinguishes organisms from inorganic objects. Something about growth and metabolism, reproduction and adaptation to the environment..

Wikipedia defines life as "Any contiguous living system is called an organism."

The free Dictionary and Merriam Webster in part had the most hilarious definition for life, "The property or quality that distinguishes living organisms from dead."

Definitions vary depending on the agenda of the person that is defining it. For instance, the biologist may have a difficult time defining life in order to put forth their particular theory on abiogenesis.

Compilation of New Data in Physics and Cosmology and Its Application in Support of the Theory of Abiogenesis. - by Biologist Nasif Nahle, with the Council Leadership/Researcher Biology Cabinet Organization, New Braunfels, Tx.

Definition of Life:

We do not have a direct definition of life, but from direct and indirect observations of the thermal state of the living structures, we can say that:

Life is a delay of the spontaneous diffusion or dispersion of the internal energy of the biomolecules towards more potential microstates.

Huh?

The first response was "everything is alive." Really? I suppose I shouldn't have flushed my goldfish or buried my dog then. I'm not even going to get into that ridiculous proposition.

Even a small child understands that there is a difference between a living object such as a goldfish, and a non-living object such as a rock. However, for purposes of a scientific discussion we need to define our term precisely, unambiguously, and in a non-synonymous and non-circular manner.

Science studies objects, that which has physical presence, that which exists; object with location. Life is the set of all living objects. So we need to define alive or living. If there are any exceptions, we need to start over with a different definition.

The second definition, "To be alive is to have life..." is not really a definition it is circular. The definition contains a variation of the term being defined. What have we learned?

What about reproduction? The very young and very old do not reproduce. Are they not alive?

So what distinguishes living objects from non-living objects? Non-living objects are different than living objects in their inability to move by themselves against gravity. This is the one criteria common to all forms of life on earth.

Non-living things move away from or towards each other due to collisions or because of a stronger gravitational pull from elsewhere. Gravity is the attraction between two or more objects.

An object is static conceptually, but because all things are in motion on an atomic level, there is a difference between living and non-living objects in regards to motion and the definition has to account for this difference. Hence the rational definition for a living object: a natural object which "moves against gravity by itself."

Living things move by themselves against gravity. This movement is by way of surface to surface contact between objects just as is non-living objects.

An earthworm receives signals that there is rain above, and it moves in that direction by wiggling his body against the soil. An E-coli bacteria wiggles its flagellum against blood platelets and moves upstream. A virus just gets pushed along by the flow of blood. That ends the on-going debate about whether or not a virus is alive.

Mountains form when tectonic plates collide and volcanoes erupt. Plates collide and volcanoes erupt because of moving magma. Magma is formed because pressure creates temperatures that melt rock. Pressure is caused by gravity pulling everything towards the center of the earth. Ergo, gravity causes mountains. Mountains are not moving against gravity, but because of it!

Gases move in water in different ways related to specific gravity of the water, solubility of a particular gas, and other things such as temperature. Any motion of gas in water can be attributed to surface to surface contact between water molecules and gas molecules. Neither a gas molecule nor a water molecule moves by itself against gravity.

A monkey moves against gravity when it climbs a tree and swings on a vine. If he drops a banana, it has fallen and can't get up. If the monkey falls out of a tree, he can get up. Whether the monkey has eaten a banana or anything else in a very long time, he can still move against gravity by himself. Metabolism takes place whether he eats or not and will continue until all the body process break down and the monkey dies. There are many contributing factors to how and why a monkey moves. No one thing determines his movement alone.

The definition 'moves by itself against gravity' is the one thing common to all living objects, without exception. It is what distinguishes living objects from non-living objects - without exception.

Chapter Eighteen – The Sense Of Touch

Humans really only have one sense, and that is touch. Surface to surface contact between molecules of air and the sensory apparatus 'ear' stimulate electro-chemical reactions in the auditory cortex resulting in what is known as hearing.

Surface to surface contact between molecules of various substances and compounds with two patches of sensory cells in the nose relay information by olfactory nerves to an area of brain that converts those impulses into something called smell. Surface to surface contact between food and the tongue produces a brain response called taste, and light touched rods and cones on the retina in the back of the eye result in something called vision which is experienced in the visual cortex of the brain.

The sense called touch is surface to surface contact between objects and receptors on the skin, muscles, bones, joints, and organs. There are chemical, temperature, mechanical, position, and pain receptors that relay information to the parietal lobe and the cerebral cortex.

Taste combines tactile, auditory, and chemical cues. Some ingenuous scientists at the University of Tsukuba, Japan have created a virtual food taster using a thin film force sensor, a small microphone to record jawbone and teeth vibrations, and lipid and polymer membranes to detect chemicals representing sweet, sour, bitter, salty, and savory.

Chapter Nineteen – Temperature, What Is It?

What is boiling water?

To understand the question, we need to define the key word 'boil' – verb, motion/vibration of atoms up to some predefined standard.

Atoms of water are constantly vibrating. When the atoms of water vibrate to a level which we define, we call this event boiling water. It is a heat source exciting the molecules of water which causes the water to boil.

There is no such thing as 'heat' in the universe. Like energy, heat is a dynamic concept that only refers to atomic motion.

In the 1800's it was thought that heat was like a substance, a type of energy. Of course energy is also a dynamic concept having to do with the motion of matter over time.

According to Wikipedia, James Clerk Maxwell proposed in 1871 that:

Heat is 'something' which may be transferred from one body to another, according to the second law of thermodynamics. It is a measurable quantity, and thus treated mathematically. It cannot be treated as a substance, because it may be transformed into something that is not a substance, e.g., mechanical work. Heat is one of the forms of energy.

Maxwell pointed out in his book The Theory of Heat that in the 1800's the word "caloric" was introduced to signify heat as a measurable quantity. So long as the word denoted no more than this, it might be usefully employed, but the form of the word accommodated itself to the tendency of the chemists of that time to seek for new 'imponderable substances,' so that the word caloric came to denote not merely heat, but heat as an indestructible, imponderable fluid..." - The Theory of Heat, J; Clerk Maxwell

There is no heat "source" anywhere in the universe. There are atoms changing location.

Does nature know to continue this process until the water boils? No!

Cold water coming out of your faucet is boiling in some context other than when you boil water to cook your macaroni and cheese. The ice in your refrigerator is boiling in another context of atomic vibration that we can define.

Altitude lowers the boiling point of water because of lower atmospheric pressure. Water will boil without releasing heat in a vacuum of extremely low pressure. And in this situation the atoms vibrate much less than water boiling at 100 C.

It all depends on what you define boil to mean in terms of the motion of atoms. Hot, cold, and boil are observer-dependent concepts, that can have different definitions in different contexts. These contexts have nothing to do with Mother Nature!

She is only concerned with now, and now can be represented by a single frame of a film showing every single object in the universe. Running the film, one would see objects continually changing locations.

As far as Mother Nature is concerned, the atoms of water only have location. Because we have memory and compare various locations of atoms, or object motion (that is, rates of change that we conceive of and measure as temperature) it satisfies our definition of boil.

In the context of cooking macaroni and cheese, that context of boil has to satisfy our sensory mechanism when we determine al dente and when we put the macaroni in our mouth. It is not the same thing in the context of water boiling in a vacuum under low pressure. We would not sense our meal as hot in that particular case.

Heat, cold, or boil do not exist in the universe. That is only the conceptual nonsense that hairless apes have invented. To coin a friend, "There is only atomic or object motion. The rate of change of atomic motion that you conceive as hot, an alien may conceive as extremely cold!"

The idea of heat, or energy, as a dynamic concept is a contrivance of man. Thermodynamics and entropy is nonsense put forth by the mainstream Phizwhiz, and is ignored by nature.

Temperature is a name we give to our conception of the rate of the perpetual change of the location of atoms and objects.

Chapter Twenty - Dimensions

I asked this question in a forum once. A number of us discussed it at length:

If one is traveling in a spaceship with unlimited fuel, could one ever reach the edge of the universe?

(It is important to be clear that the question about a sentient being [you, me, "one"] reaching that edge is not solely, or implicitly, about its existence. Hopefully, 'The' Universe exists whether or not there is a sentient being to observe it.)

Someone answered, "As one travels towards the edge of the universe "the further away one gets, the faster the universe is expanding. (Cosmologists talk about an expanding universe.) At some point the spaceship would have to "out accelerate" this expansion to be able to reach the edge (if there is one). Of course, this depends upon you being in the universe in the first place, and if you are able to go faster than light."

Hawking and Penrose describe a model where there not only is real time- which had a beginning at the big bang, but there is imaginary time- which has no beginning or end. Can you imagine that?!? From everything forever.com

According to Stephen Hawking:

"The boundary condition of the universe is that it has no boundary. The universe would be completely self-contained and not affected by anything outside itself. It would neither be created nor destroyed. It would just BE."

We are reminded that this place outside of the universe is where men put God. We supposed there could be something outside the universe, because after all...the universe needs something to expand into? Hmmm...does it or, doesn't it? We have to also consider that if there is no boundary there can not be a center. You could travel, and travel, and travel (are we there yet?) and finally arrive back to where you started.

Jim imagines... that "traveling through space" has the universe wrapping around it, almost as if the universe spins around magnetic field lines drawn along the direction of travel. Every moving thing in the universe would be the "center".

Theoretical physicists tell us that the laws of physics vary across the universe, thank you very much! If we took off to the edge of the observable universe now, when it is currently 47 billion light years away, by the time we got to where that edge was it will have been expanding for 47 billion years and would be even further away.

From wikisource.org:

"Space does have an inherent anti-gravity force, called the Dark Energy. Space has a negative, Lobachevskian curvature. Therefore space has no boundary."

Another person said that it doesn't matter if there is an edge because the universe is an expression of spacetime. Time comes into existence when something changes. Your presence in space would change things. You effectively create time. Wherever you are would be "in the universe." Therefore as you got to the edge of the universe, time and space would unfold before you.

We imagined a 2D universe shaped like a ball with an ant walking around on the outside of it. We also imagined a jar shaped and a Mobius strip shaped universe. An odd thought is "if the universe is self-contained, then everything that has happened, is happening, and is going to happen is all there- in one place."

The universe has been described in this way: We (everything in the universe) are actually on the surface of an expanding beach ball. No matter how fast or how far one goes one can never reach the edge. There is none. You could go in one direction long enough to end up where you started out. Everything is moving away from everything else and the further out the faster things are moving. Out, away from the not center, towards the no boundary.

Of course, according to Calabi Yau, the real universe is "4-D expanding through an 11-D manifold- and all of this makes sense when you understand: It's a wrapped, temporal-volumetric expansion."

Finally, and perhaps most importantly, we see that one could exclusively study for their entire lifetime and barely scratch the surface or be unable to explain what they know to other people. One person said he would "leave this to the guys with bigger telescopes."

The atlas of the universe.com website has an illustration of "the universe" and you can see the "edge" for yourself.

What is on the other side of that edge?

In The Universe In A Nutshell, Stephen Hawking has us consider that there is no edge, just as there is no edge to the expanding beach ball. Huh?

It is difficult to understand the 'incomprehensibleness' of the cosmos. We are not consciously aware of all the many dimensions that inform our existence. It takes math, which uses imaginary numbers and is difficult for many to relate to. We are told that the 11 dimensions (of string theory) can explain a lot but 26 dimensions have been considered. Perhaps there are infinite dimensions.

It is not easy to understand the fourth dimension (spacetime) and it gets increasingly more difficult to understand higher dimensions. What about the idea of nothing as opposed to everything? We consider that it may be pointless to refer to something 'outside' of the universe, since by definition the universe is all there is. In order to understand theoretical physicists are following the expansion backwards in time towards the Big Bang.

"We must try to understand the beginning of the universe on the basis of science. It may be a task beyond our powers, but we should at least make the attempt. " - Stephen Hawking

The laws of physics which apply to the universe that we know are different at and before the Big Bang. Small stuff behaves differently than large stuff. Scientists are attempting to combine non-quantum general relativity which gets us in the neighborhood of the big bang, with quantum physics, which seeks to explain (among other things) that period of time known as the Planck Epoch.

Since, we are told, space is inseparable from time, let's define time before catching up on our discussion. Physical time is the measurement of motion. Movement through space- we've already said that. Now let's break it down further. Our base unit, the second, is derived from the decay of cesium:

From a HubPages blog by Shivra:

"The second is now established as 9,192,631,770 periods of the radiation corresponding to the transition between the two hyperfine levels of the ground state of caesium-133."

The units of time extend forward and backward from the second. The largest unit of time is the exasecond 10 to the 18th power S(seconds). That is about twice the age of the universe at 32 billion years. There is the millennium at 1000 years and the indiction at a 15 year cycle and the lustrum, or pentad, weighing in at 5 years. And what would a time scale be without the fortnight (2 weeks)? Heading towards the other end of the scale we find our old friends the millisecond and the nanosecond along with the not so memorable femtosecond at 10 to the neg 15th S.

And the winner for the smallest unit of time that we can measure is...the attosecond at 10 to the negative 18 S. Attoboy attosecond! After that, we have the zeptosecond at 10-21S, and the yacosecond 10-24S. But the time period that physicists are mostly interested in, is the Plank Epoch.

Wikipedia says, "In physical cosmology, the Planck epoch (or Planck era), named after Max Planck, is the earliest period of time in the history of the universe, from zero to approximately 10^{-43} seconds (Planck time)."

Getting back to our discussion, Stephen Hawking in The Universe In A Nutshell talks about imaginary time (which is made of imaginary numbers) to allow us to describe mathematically what we can not imagine...like infinity. "We can't have an imaginary number of oranges" he says, "but we can find which mathematical models best describe the universe that we live in." Imaginary time describes the past as well as the future and predicts effects we have already experienced, as well as those we have yet to measure.

Einstein's General Theory of Relativity combines the three directions of space (X,Y,Z) with time into a four dimensional space-time. Since imaginary time is at right angles to real time it acts like a fourth dimension in space.

Hawking says we can look at imaginary space-time as a sphere with imaginary time as degrees of latitude or degrees of longitude. With imaginary time (IT) as degrees of latitude, time begins at the South Pole. As one moves north the circles of latitude get larger corresponding to the expansion of the universe. It is biggest at the equator and gets smaller and smaller until it resolves into a single point at the North Pole. The origin of the universe could be any one point (or none at all).

We can also look at imaginary space time as a sphere with longitudinal lines. All the lines meet at the poles and so time stands still at these points (since an increase in IT leaves you at the same place). This is similar to the event horizon of a black hole where real time stands still.

Hawking, Penrose, and others discuss multiple histories. There is some connection between heat (and entropy) and quantum gravity. This also needs to be discussed (but not now). There are theories which suggest quantum gravity exhibits holography. Imagine three dimensions being presented on a 2-d surface where any part of the holographic image contains the entire image (but from a different angle). Now expand that idea to include 11 or 26 dimensions.

"It seems we may live on a 3-brane--a four dimensional (three space + one time) surface that is the boundary of a five dimensional region, with the remaining three dimensions curled up very small. The state of the world on a brane encodes what is happening in the five dimensional region."-Stephen Hawking

"We can't find the boundary, because we are the boundary. Talk about living on the edge! " - Monk E. Mind

"The real universe is "4-D expanding through an 11-D manifold"- and all of this makes sense when you understand: "It's a wrapped, temporal-volumetric expansion." - Cyberia

"This dimension is the surface tension between higher and lower dimensions (not!)." - Monk E. Mind

I like to (mis)use the terms interference wave pattern and surface tension. When looking at a wooden table we see where the molecules in the surface of the table meet with the molecules of air surrounding it. We see the 'interference wave pattern' not the air or the wood molecules. Now extend the analogy to dimensions. Where two dimensions intersect another dimension (a surface tension) 'arises'. Each dimension itself the surface tension between two others.

If we step through the transition of a dimensionless point into a three dimensional sphere we see that each dimension is 'contained' within the other. A point merges into other points, which come together to form a line, which join with other lines to form a plane, which may be the surface of a sphere.

The line is the surface tension between the point and the plane. The plane is the surface tension between the sphere and the line. The sphere is the surface tension between the plane and space-time. Now replace surface tension with the term boundary.

The three spatial directions (XYZ) and Time (the fourth dimension) combine to form space-time. Up/down, left/right and forward/back joins with time, which is movement through space, along with other dimensions interfaces with both the macro scale of planets and galaxies, and with the nowuseeit/nowudon't micro scale of Quantum incomprehensibleness.

I'd say Quantum strangeness, but that term is already in use.

I'd say weirdness, but that too is taken.

So I use this term, Quantum Incomprehensibleness (QI). From MonkeyPedia:

Quantum Incomprehensibleness (QI) is the attribute of being very, very awesome and unfortunately unfathomable or very, very difficult for the layperson to understand (It's difficult enough for the scientist).

It's the stuff that's so strange that, well, there are just no words to describe them. There are many mathematical formulas and theories and hypotheses which most monkey minds have difficulty understanding. There are many attempts to describe this Quantum Incomprehensibleness.

According to wikipedia, the German Philosopher Immanuel Kant's Antinomy of Pure Reason) asks:

"If the universe was created, why did it take an infinite time of waiting before the creation? If the universe has always existed, why hasn't everything that is going to happen, not happened already? Why hasn't the universe reached thermal equilibrium?"

For convenience's sake, calculations, equations, theories, and approximations often use infinite series, unbounded functions, etc., and may involve infinite quantities. Physicists, however, require that the end result be physically meaningful. In quantum field theory infinities arise which need to be interpreted in such a way as to lead to a physically meaningful result, a process called renormalization. Sure, sure.

However, there are some theoretical circumstances where the end result is infinity. One example is the singularity in the description of black holes. Some solutions of the equations of the general theory of relativity allow for finite mass distributions of zero size, and thus infinite density. This is an example of what is called a mathematical singularity, or, a point where a physical theory breaks down. This does not necessarily mean that physical infinities exist; it may mean simply that the theory is incapable of describing the situation properly.

Lawrence Krauss, in A Universe from Nothing, says that he lies when he uses words that only mathematics can describe.

What is nothing according to wiki?

"In physics, the word nothing is not used in any technical sense. A region of space is called a vacuum if it does not contain any matter, though it can contain physical fields. In fact, it is practically impossible to construct a region of space that contains no matter or fields, since gravity cannot be blocked and all objects at a non-zero temperature radiate electromagnetically. However, even if such a region existed, it could still not be referred to as "nothing", since it has properties and a measurable existence as part of the quantum-mechanical vacuum."

According to Lawrence Krauss, in A Universe from Nothing:

"Nothing at a quantum scale is something." Looking at a proton we find that most of the mass of the proton is in the empty space between quarks. Dark Matter. 90 % of the mass of galaxies and clusters is dark matter (and dark energy). 90% of YOU is empty space. Take away the cosmic microwave background, the galaxies, the stars and all the 'stuff' of the universe, you still have a "boiling, bubbling brew of virtual particles popping in and out of existence in a time scale so small you can't see them."

So in Krauss' universe, nothing is something. Galaxies expand forever. We live in a flat, curved, open universe that will expand forever at an accelerating rate and is filled with mostly nothing. And in a flat universe, with the ratio of normal matter to dark matter balancing out (where energy = zero) the universe can begin from nothing.

So...the universe can begin from nothing and can expand forever?

Is there an edge to the universe? Could one ever reach it if there was? Lawrence Krauss suggests that as galaxies move far away from each other, and at an accelerated rate (possibly faster than light) there may be a time in man's future (perhaps 100 billion years from now) where galaxies are so far apart from each other that scientists at that time could reasonably calculate there are no other galaxies..."and they will be wrong."

"There is no edge...there are no boundaries, no limits to the intellect." - Monk E. Mind

Chapter Twenty One - Dimensions of Reality

Length, Width, & Height

We mostly raised a lot of questions in Part One. Let's see if we can answer some of them.

Is there an edge to the universe?

We never defined universe. We should have. It is unscientific to make a presentation, or involve ourselves in a scientific discussion without first defining our key terms.

The universe is a concept comprised of matter and space. Matter is the set of all objects. Objects are that which have shape. Space is that which lacks shape. Shape is defined as a border, or boundary which separates that which is inside the border, from that which is outside, or surrounding the border. Shape is the one intrinsic property common to all objects.

Science studies, and seeks to explain objects. Shapeless, borderless, space is a concept we use to relate all objects. How can there be an edge to borderless space?

All objects are finite. If space were an object, it could not be finite. If space where some 'thing' it would have to be a solid block of matter. Motion would be then impossible as there would be nowhere to displace objects. If space were an object, with a border, what would be outside of the border? Space is a where, not a what!

Traveling through space.

If space is a concept, how can one travel through it? Space is not a medium. There is no fabric called space. One can not bend, or warp, or ripple 'it', and one can not travel through 'it'. All motion is the result of surface to surface contact between objects. On earth, one travels by pushing their feet against the ground. On the trip to the moon, the rocket exhaust pushes against the rocket.

What's on the other side of the edge of the universe?

In Part One, we looked at a map of 'the universe'. We saw that there was an edge to it. What was on the other side of that edge, or boundary? More space? No! Space is border-less. Only objects have borders.

The incomprehensibleness of the cosmos, many dimensions, and imaginary numbers...

Theoretical physicists would like us to believe that the universe is too complicated for the poor, uneducated layperson to understand. It's such a strange and weird place that it is too incomprehensible for any human to understand. We are told to imagine 11, or 26, or even an infinite number of dimensions. We are asked to believe in spacetime and told that we can understand spacetime if we take the three dimensions of reality, length, width, and height, then superimpose a time line with imaginary numbers on it. Don't buy it!

Man may be limited in his sensory capabilities, but he is unlimited in his intellect...the ability to conceive of concepts. One should NOT be expected to 'understand' the nonsensical. Infinite, four, or 11 dimensional realities are irrational, and can make no sense. There are only the three orthogonal dimensions of length, width, and height. If anyone thinks that there is a fourth dimension called time, please draw a picture of it at right angles to the three dimensions of reality and post it for all to see!

If someone tells you that they understand something, then they can explain it to you. If they can visualize it, then they can illustrate it, or, create a mock up, or sculpture of it. No one has ever been able to do this with time or any other than the three dimensions (collectively).

Time is a concept. 'It' does not exist. 'It' has no dimension to 'it'. No length, or width, or height! 'It' is not a physical substance that can dilate, and neither can anyone travel through 'it', or go backwards in 'it'. Time is an arbitrary construct man devised to help us order our lives, based on the motion of the earth and the sun. As discussed in part one, we arbitrarily use the vibration of the cesium atom for our second, in order to calibrate our clocks. The cesium atom exists, as it has the three dimensions of reality and has location in relation to all other objects. Vibration does not exist, as it is a relation between a steady state and an excited state of the atom (also called quantum jumping). It is a phenomena, a concept.

Did you know that a day on Venus is longer than its year? Just because Venus' rotation on its axis is of greater duration than its orbit around the sun, does not mean that some 'thing' called time moves differently relative to the earth. All motion is the relation of an object to all other objects. Any event here on earth happens at the same 'time' in relation to everywhere.

Monk E Mind's analogy of interference waves and surface tension

Although it sounds nice, maybe even a bit poetic, it has nothing to do with reality. A point has no dimensions, so it can not be part of a line. A point is but an arbitrary location on a line, inseparable from the line. If a line is only one dimension that of length, then how does it become the edge of a plane? The point, the line, and the plane, are non-existent. Press a pencil to paper, and you get a point, or a dot of graphite. The graphite exists, the point does not. A piece of paper may exist, but not a plane. There is thickness to the paper, so it is a three dimensional object, not two. The square does not exist, nor the circle, nor the triangle. These all have shape, therefore they are objects. But they do not have location, therefore they do not exist.

Infinite density?

Infinities are irrational. No such 'thing' exists in reality. An object is finite by definition. A singularity (meaning singular; that is, one [object]) can not be infinite. That is a contradiction. Nor can an object have an infinite attribute such as density. Can you conceptualize infinite compactness?

What is nothing?

Zero dimensions. No height, no length, no width, no substance, no material, no physical presence. The only word in the human language that can only be defined by its negative predicate. Another word we use for nothing is void, or space.

The universe can begin from nothing

Beginning of the universe. Based on our definition of universe, that is an irrational proposal. Did matter have a beginning? No. All matter is eternal. If anyone thinks differently, then they need to explain how zero dimensions becomes three dimensions in one frame of the universal movie.

Is there an edge to the universe?

Impossible! The universe is a concept, and as such, it does not exist!

Chapter Twenty Two - The Three Dementia of Geometry

If you haven't read Edwin Abbott's novella, Flatlander, you can download an e-book copy of it from Amazon for 99 cents. It's very entertaining. Although, being written in the 1800's, it may take some effort to wade through the archaic but poetic language.

Flatlander is a character that lives in a two dimensional (plane) world, and he describes in great detail what it is like to those who reside in space land (where we fortunate three dimensional folk live).

Of course it is very difficult to discriminate a line from a square, or a circle, or a polygon. The lower caste, and less educated, must resolve to touching in order to distinguish the difference. One may forgo the touchy feely side of Flatland if they have been formerly trained in the University to sense shape, or, if one is of the nobility, like a hexagon or, the circle (the priest), who have the highly trained ability of sight.

Flatlander is a square (a lawyer). He lives in a pentagon home, as do all residents of Flatland. This is decreed by law because of the danger that a sharp edged domicile may pose to the unwary isosceles triangle, or line that might inadvertently wander into it.

Women are like unto lines (read needles), and as such are very dangerous when not standing sideways, as their pointed end has pierced many an inhabitant of Flatland resulting in injury or death.

Flatlander has been brought to the world of three dimensions, and upon his return to Flatland discovers that no one will believe, or in fact, can understand him. To them he must appear quite mad. He dreams of a visit to Line land and is unable to explain to the Line Lander from whence he came.

Presumably, the second dimension is incomprehensible to the Line Lander, as the third dimension is incomprehensible to the

Flatlander, and just as the fourth dimension is to the Space Lander.

Exasperated by his attempts to inform the Flatlander about his three dimensionality the sphere says:

"Why will you refuse to listen to reason? I had hoped to find in you-as being a man of sense and an accomplished mathematician-a fit apostle for the gospel of the Three Dimensions..."

Eventually Flatlander is imprisoned for talking about his experience in a higher dimension (preaching a gospel other than the Gospel of Flatland).

So it is also with the geometer and the relativist who seeks disciples for the gospel of Four Dimensions.

However, what one quickly discovers, is that the analogies a Flatlander must use to describe his two dimensional world are actually 3D scenarios. For how can one touch, or sense, or see anything in a land with no height? The Flatlander attempts to describe his world like this:

Place a penny on a table. Viewing from above, one sees a circle. As one bends over towards the edge of the table the circle becomes an oval, and when viewed edge-on it is as a line.

Well, it is not quite a line, is it? What one sees is a rectangular shape. The height is far less than the length, but it is clearly not a one dimensional line with only length.

Like the fantasy world of Flatland, geometers in the real world are faced with a similar (but self-imposed) dilemma. Mathematicians must rely on analogies to explain higher dimensions. BUT… this is not because of a limit to our neural wiring, or because we lack imagination. It is because higher dimensions are nonsensical.

They can not be visualized, illustrated, or sculpted from clay.

Riemann's triangle and Hinton's Tessaract are abstract concepts with no corollary in reality. Time as a fourth dimension running perpendicular to length, width, and height, is a product of overactive imaginations in an attempt to reconcile failed theories.

These silly concepts are based upon geometry, a branch of mathematics which seeks to relate shapes, and sizes, but does not describe reality because it is corrupt at its core. Although Tessaract and tribars make wonderful art posters, they can serve no purpose in science, except perhaps to aid neuroscientists and psychiatrists in their quest to cure dementia. Geometers such as Euclid and Riemannm are currently drawing square circles with their feces on the walls of their padded cells...

Three basic units of geometry are point, line, and plane. Take a closer look at these, and you will see that these terms are left undefined, are defined ambiguously, and used inconsistently.

Depending on its use, the point can be a number of things. It can be a dimensionless object with only location, an infinite set of 'dots' in a row or, it may be represented by an ordered pair.

"The word 'point' is often left undefined in geometry texts. It is pretty easy for us to conceptualize a point, but it is quite difficult to define exactly." From Ask Dr. Math: mathforum.org.

Well, that's not good, because a point makes up the line, the line is crucial for the plane, and the plane is an integral part of the solid. Perhaps this is why Relativism, Quantum Mechanics, and all of particle physics, which depend on geometry, fail miserably at explaining our physical existence.

How does one build lines from points, planes from lines, and solids from planes if the point is zero dimensional? Is the point an infinitesimal dot or zero dimensional! Put your pencil point to paper. What do you have? A dot. This is not just the two dimensional object one sees with the naked eye. Under a powerful magnifying glass one can see that it has height, as well as, length and width.

Now draw a line. How can there be an infinite number of these dots between the two end points? Of course, there can not. Also, if the line is defined as the infinite number of locations between the endpoints, what happens to the line when you remove the endpoints? If the point is but an abstract concept 'location' then one can not build a line from these imaginary 'objects' and one can not then build a plane of that, let alone a solid.

Another definition of a point is where two lines meet. We are also told that a line requires two end points. Two lines for a point and two points for a line! That is circular!

Euclid defined it like this: "A line is breadthless length." Since his time, mathematicians have come up with a lot of different definitions for the line, from an intersection of two planes, to an infinite number of points, the shortest distance between two points, and the result of an equation.

Why so many definitions for the point and the line? Why do mathematicians confuse dimensions with coordinates and vectors, and fantasy with reality? They do this because they refuse to define their terms unambiguously and then use them consistently. Perhaps they are confused, or perhaps they understand that their theories fail if they adhere to precise, consistent definitions. And perhaps mathematics has nothing to do with reality.

Riemann takes a triangle, places it on a sphere, and notes that its 'angles' are greater than 180 degrees. Yet, the angles are contingent upon a plane. A plane is two dimensional, but when it is 'bent' to conform to the sphere it is no longer 2D, or a plane as it has encroached upon the world of the spacelanders. It is now 3D.

A tessaract is illustrated as a cube within a cube, with the inner cube's corners connected to the outer cube's corners. Note that the angles are NOT 90 degrees! Dimensions require orthogonality. All dimensions of an object must face outward at 90 degrees to each other.

The three dimensions of reality (L, W, & H) do this. Now add the fourth dimension of time as relativists are wont to do. Where does this fit? And how could time be a dimension anyways? Time is a concept. Time is motion plus memory. Surely the objects of reality do not depend on the observation and memory of a sentient being to exist.

Dimension: any of three mutually perpendicular directions in which an object faces or points simultaneously

An object is that which has shape. Geometry is the study of shapes. Do the three dementia of the geometers, the point, the line, and the plane, rationally describe objects? Maybe we could answer this question if they would settle on unambiguous, non-contradictory, non-synonymous definitions and then use them consistently.

So which makes sense to you?

'The three dimensions of reality' or 'The three dementia of geometry.'

Chapter Twenty Three – Time Part One

Anybody Have The Time?

A friend sent me this email a few years ago before I had looked at 'time' using the rational scientific method. Here it is along with my first response.

Time Fractals and repeating history

"It's interesting. For a long time, I've felt that time occurs in cycles and has a certain spectral nature, based upon my background in physics, wave theory, and wave form analysis. Now I think of time fractals and time harmonics. I think I really began to become aware of this in 2008. Recall the expression, 'History repeats itself?' I've often 'felt' future events, but now I'm looking for more quantifiable methods to my madness through mathematics, physics and systems of probabilities.

We always say, 'Timing is everything' or 'There is a season for everything.' I think this is our innate understanding of the cyclical nature of time and events.

Sometimes we are 'in the flow' and everything just clicks into place; at other times, we have 'reverse Midas touch' where everything we touch turns to crap instead of gold.

If time is fractal and occurs as waves, then it follows the basic laws of electromagnetic wave theory and can be represented through vector algebra and Fibonacci sequences which appear to occur throughout nature.

Note: Technical analysis of the financial markets relies heavily on various forms of fractal theory, and my experience is that it is highly reliable.

Simply put, if two wave forms of equal and opposite polarity or phase relationship and identical magnitude are mixed, the 'vector sum' becomes zero. If they are in phase, then the vector sum is equal to the sum of their individual magnitudes and the signal is stronger.

Magnetic or any other energy field expressed in a 2 dimensional expression such as a sine wave, is really expressing the expansion and collapse of an energy field in a rhythmic sequence.

The word sequence can be misleading, because evidence suggests that time as we understand it with the physical senses is an illusion. The reality would be most likely that time is an eternal moment, expressed in infinite ways.

We know that we can never do something yesterday and we can't do it tomorrow. We know we can't do an action or 'be' anything 3 seconds ago or 3 seconds into the future, because as soon as we are aware of doing/being it, it is 'now'.

Place a pencil on your desk. Attempt to 'feel' the now moment...or the sense that you are perfectly balanced between what just happened and what is about to happen.

Try to pick up the pencil 1 or 3 seconds in the past. You can't because the past doesn't exist. It's a psychological construct.

Now, try to pick up the pencil 1 or 3 seconds in the future. (You'll most likely count to 3 and pick up the pencil - almost everyone does.)

Notice that as soon as you 'touch' the pencil, it is no longer 3 seconds into the future...it is now. 'Lather, rinse, repeat' as the shampoo instructions would say, until you understand truly what you just experienced. Some will 'get it'...most will either take more time or won't.

I came up with this idea quite spontaneously years ago. I shared with my kids, and when they understood it was amazing. (They would ask, 'What time is it?' and I would always answer, 'Right now.') Have you noticed that it's never yesterday and it's never tomorrow. It's always now. (I guess Nike says it best with their slogan.)

If it is always now, then how long does now last? Did it have a beginning or an end? It would mean that now lasts forever. Maybe this is the meaning behind the words, 'I am the alpha and the omega.'

Imagine an 'infinite moment', expressed like a kaleidoscope of patterns that, although the general pattern is repeated, the exact expression of that pattern never repeats.

What if time and history are cyclical repeating mega patterns, but human choice determines the infinite variety in how we express it in smaller patterns?

Would this explain why old prophets could understand certain things about our future, a future that is now beginning to unfold and that is about to accelerate with greater intensity?

I think everyone alive today is very special. We live in special times...interesting, chaotic, frightening...but fascinating to experience. But there 'is' order in the seeming disorder and chaos.

I believe that each of us, in our own private way, has unique and special things to do in our life during these times.

Endings become beginnings. That is exciting to me.

By the way...what time is it?"

"Yesterday is history, tomorrow is a mystery, and today is a gift, and that is why we call it the present." - Eleanor Roosevelt

Me: "Your ideas are interesting, but I am having some problems relating in a couple of areas. I learned a little about the propagation of electromagnetic waves when I was getting a degree in electrical engineering, but I am a bit rusty - and I'm not sure I understand what I know about it!

From Wikipedia:

In mathematics and science, a wave is a periodic disturbance in space and time, possibly transferring energy to or through a spacetime region. A wave equation describes how the disturbance proceeds over time. The form of this equation varies depending on the type of wave: ocean waves, sound waves, electromagnetic waves.....matter moving through space over time, or space-time.

Also in Wikipedia, speaking of a sinusoidal wave:

'The sine function is periodic, so the sine wave or sinusoid has a wavelength in space and a period in time.'

So the problem that I see, is the use of the word 'wave' in understanding time.

If a wave is the motion of physical matter in space over time, how can you call time a wave? It would be like saying time is matter. I understand time on the macro level to be movement through space and, if I understand correctly, in quantum physics (space)time is referred to as being the 4th dimension.

How does one use time to explain time (circular reasoning), how can time be matter, and how can a dimension be either of these? Finally, if time is a form of matter, then the old saying which goes something like this: 'Time is an invention of our mind, so everything doesn't happen all at once' or, in your words 'evidence suggests that time, as we understand it with the physical senses, is an illusion.' would appear invalid.

As for fractals, Wiki says this:

'A fractal is a rough or fragmented geometric shape...'

and this:

'Natural objects that are approximated by fractals to a degree include clouds, mountain ranges, lightning bolts, coastlines, snow flakes, various vegetables (cauliflower and broccoli), and animal coloration patterns."

So, I run into the same problem I had with using the term waves, in that fractals are defining shapes, objects, clouds, mountains, i.e., things.

Chapter Twenty Four – Time Part Two

If I Could Save Time in a Bottle

I had another discussion with a man that claimed to be the lead scientist who designs the atomic clocks on the GPS satellites. Afterwards, I began to take a much closer look at this concept of 'time'.

GPS Guy: "Defining time is a live field of physics. There are experiments being considered to see if time is in fact something inherent to the structure of the universe."

ME: The question of time can be solved here and now rationally.

GPS Guy: "One of these observable things we would still have to account for is that the passage of time does in actual fact slow down or speed up depending upon the position you occupy in the universe. We have measured this over and over again."

Me: Time cannot pass, slow down, speed up, or be measured.

GPS Guy: "These are the principles you have to disprove in order to show that relativity theory is somehow mistaken. Good luck to you in doing so."

Me: Sorry, but proof has nothing to do with the Scientific Method. Theory does not prove, it explains. However, because relativity has loosely defined terms, brought common conversational language into science, and uses those terms ambiguously, Theory of Relativity is not falsifiable which means it is not science! [according to his method of scientific inquiry] The Rational Scientific Method only rationally explains, and does not concern itself with true or false, or other matters for religion.

GPS Guy: "Summary: If you're ticked about relativity theory not defining time, then you're ticked at all of physics and not relativity."

Me: Being ticked at relativity does not invalidate it, nor does loving it validate relativity.

GPS Guy: "Also, if I may, I'd like to ask a question... What sort of time dilation makes you upset: high speed or proximity to mass, or both? If one, the other, or both, please explain why this is so. Although they both can be understood under General Relativity, it is actually quite a bit easier to talk with non-physicists about time dilation in Special Relativity."

Me: Neither make me upset. Time has nothing to do with dilation, speed, or mass. I am sure that it is easier to talk to non-physicists, since there are varying opinions. However, opinions have nothing to do with physics in general or time in particular, and the average Joe doesn't understand this.

Thank you for taking your time to explain some things. Sorry, I don't have time to answer in greater detail.

"Insofar as mathematics is true, it does not describe the real world. Insofar as it describes the real world, it is not true." - Einstein

GPS Guy: "1) The passage of time can be measured by accurate well-calibrated clocks."

Me: Time is a man made concept. It is an object at 2 or more locations plus an observer's memory. Nature doesn't remember anything. Without memory, there is no time.

GPS Guy: "2) The laws of physics should be the same for all observers, regardless as to their frame of reference (the principle of relativity)."

Me: Nature doesn't recognize laws. Again, it is a man-made concept.

GPS Guy: "3) Measuring the speed of light will always produce a constant value, regardless of the frame of reference of an observer."

Me: Science, especially physics, is observer-independent.

GPS Guy: "By calculating both the initial vertical and horizontal velocities of a projectile, we can very successfully predict where said projectile will fall."

Me: We can do this without math too: have you ever played horseshoes?

GPS Guy: "Your motion through space effects time."

Me: A clock's motion through space affects clocks.

GPS Guy: "Space and time are thus intertwined."

Me: Space is nothing. Time is a concept not a thing. Nothing and a concept cannot be intertwined.

Chapter Twenty Five – Time Part Three

Does anybody really know what time it is? Does anybody really care?

Well...mathemagicians care, relativists care, and Quantum Mechanics care. Their ridiculous proposals depend on 'it'.

Is time an 'it'? Can anybody have the time? Can Jim Croce put time in a bottle? No, time is not a thing. No one can 'have' the time, and even a musician understood what physicists do not: one can't put time in a bottle. Hence, Croce sings: "If I could put time in a bottle."

Physics is about physical things. All phenomena is explained through understanding the physical objects involved.

As I began to understand in Part One, time isn't an object that can be pointed to. It is a concept involving objects, motion, and memory. It began to become clear in my conversation with GPS Guy in Part Two that time cannot pass, slow down, speed up, or be measured.

A concept is a relation between objects. Time is the relation between locations of objects and requires the memory of an observer. Since science deals with objects, and the rational scientific method removes the observer with his biases and limited sensory system, time is NOT a scientific term. As pointed out to me in yet another conversation on time:

"That's right, Monk. Time is not a scientific term. It is not a part of science because we cannot explain anything with time. Time is 100% descriptive. Time cannot tell us WHY something happened, what caused the phenomenon."

time: motion and memory

motion: two or more locations of an object

location: the set of static distances between objects

object: that which has shape

As you can see from the definitions above, time is a concept that involves an observer. It is therefore not a scientific term.

GPS Guy told me that "time dilates" and this is clearly seen in the calculus that he uses to correct for the time errors resulting from this dilation as predicted by relativity.

As I pointed out to GPS Guy, time did not dilate, what was happening that the calculus described was varying stress on caesium atoms in the atomic clock. As the clock orbits the earth, there is a varying degree of gravitational pull on the atoms as distance varies from them to the center of the earth. Relativity has nothing to do with it. Actually, the amount of correction GPS Guy uses is smaller than the claimed accuracy of the GPS, so relativity isn't relevant to GPS position anyway.

Einstein made this rational statement about time: "Time is what clocks do."

A clock measures its own movement, based upon man's arbitrary use of the relationship between the orbits and rotation of the earth and sun.

Chapter Twenty Six - Mass Confusion

After hundreds of years, the mathemagicians still don't understand what mass is. They will tell you that mass increases when an object travels near the speed of light.

Many people believe that mass is the amount, or quantity, of matter in an object. This is a rational thought.

However, depending on which relativist you are talking to, they will either tell you mass itself actually increases, or just its speed increases. Most will tell you that what is increasing is actually energy. What is energy? Relativists don't know what that is. Yet they will try to convince you that the so-called mass increase that they don't understand is really energy increase that they also do not understand.

The Relativist says that the mass of an object increases as it approaches the speed of light. They use an equation which is like the length contraction formula but they substitute relativistic mass and rest mass for the length parameters.

Anytime an equation involves infinities, you know they are using constructs which will lead to magical results. Infinities are impossible. In order for an object to travel at the speed of light, its mass would have to go to infinity. The use of infinity here is an adjective turned noun. Mathemagical physicists confuse verbs and nouns, adverbs and adjectives, and as in this case, adjectives with nouns. When the mathemagician gets tired of counting, he calls that infinity. He's really talking about incessant counting.

When discussing mass increase, Relativists cannot decide which physical interpretation to use. Some will tell you it is a problem with perception, while others will say it is a real increase in the object's substance.

A person traveling along at faster and faster speeds would notice no change in their mass.

" The idea of mass increasing with velocity leads many to believe that the internal structure of a quickly moving object is somehow altered. However, the standard view of relativity is that the internal structure of an object is always unchanged, and that the different quantities measured by different observers for energy and velocity are simply the same reality seen from different points of view." – reference.com

The person traveling and an observer, would have different experiences of inertia

" Note that the body does not actually become more massive in its proper frame, since the relativistic mass would be different for an observer in a different frame. The only mass which is observer independent is the invariant mass. When using the relativistic mass, one must always specify a velocity relative to a particular observer." - wikipedia

If mass is interpreted as a quantity of matter, then the traveler would experience a real increase in their matter:

"...as an object travels faster, its mass increases. But as the mass of an object increases, it takes more and more energy to increase its speed any further. Eventually, as the object gets close to the speed of light, it becomes so massive that no amount of energy will make it go any faster. This means that the speed of light is a universal speed limit which nothing with mass can break." – einsteinyear.org

So can we define mass unambiguously? What is the difference between invariant mass, rest mass, inertial mass, and gravitational mass? One physicist told me this:

Ron: "Rest/relativistic mass, inertial mass, and gravitational mass are all called 'mass' in common usage, but they represent three completely different concepts. The Theories of Relativity connect all three and show them to be equivalent."

Different, but equivalent...sure! Why so many versions? Why have Relativists been working on a definition since the 50's and still can't agree on a definition?

There are those that don't know what mass is:

"What is mass? What is matter? General relativity does not provide an answer; in fact, it does not describe matter at all. Einstein used to say that the left-hand side of the field equations, describing the curvature of spacetime, was granite, while the right-hand side, describing matter, was sand. Indeed, at this point we still do not know what matter and mass are." – motion mountain.com

There are those who believe that an object can have any mass at all. The following statement from professor Mark Strovink at the University of California, Berkeley:

"Virtual particles live for so short a time that Heisenberg's Uncertainty Principle allows them to have any ."

There are those who say mass and matter is the same thing. This from How Stuff Works.com:

"Mass is defined as the measure of how much matter an object or body contains – the total number of subatomic particles (electrons, protons, and neutrons) in the object."

"The principle of energy conservation implies some interesting consequences for the fundamental structure of mass. For example, if we hypothesize the existence of fundamental discrete particles of mass, with no constituent sub-particles or "degrees of freedom" that can be excited in the form of "temperature", then the collision of two such objects would necessarily be perfectly elastic. Conversely, if we observe an inelastic collision of two such fundamental objects (with no emission of energy), we would be compelled to conclude that the internal energies of the objects had somehow increased, i.e., that the energy is still present in some form. Thus we would have to believe that even these supposedly fundamental particles have some internal degrees of freedom (i.e., separate "parts") that can be excited. (Another possible outcome would be for two particles to meet and enter into a combined bound state with each other - like an electron in some orbit around a proton - so that the combination has a degree of freedom to absorb energy)."

"To most people E=mc2 is still just a bunch of letters and a number... it's a very famous equation even though people don't understand what the symbols mean...Basically it expresses the equivalence of two quantities that people previously didn't know were connected - one is energy and the other is mass. So what it tells you is that energy has mass and that mass, or matter if you like, is a form of energy, and the fact that they're connected through this equation means that you can convert the one into the other. Matter can be converted into energy and energy can be converted into matter, so that in a nutshell is what the equation is telling us." - Mathpages.com

So we have solid mass, mass particles, and mass/matter as a form of energy.

The deal is, mathemagical physicists want to move motion, measure magnitudes, and abstractly mix magical potions, rather than count units of matter. They are interested in motion, not architecture. Rather then consider shape as an objective criterion for reality, they use the subjective measurement of motion.

"When you can measure what you are speaking about and express it in numbers, you know something about it; but when you cannot measure it and when you cannot express it in numbers, your knowledge is a meager and unsatisfactory kind: it may be the beginning of knowledge, but you have scarcely, in your thoughts, advanced to the stage of science." (William Thomson on Lord Kelvin)

When the mathemagicians use the word mass, they mean inertia. They measure inertia (resistance to force) and determine mass. This is not a quantity of matter. Physicists should be counting units instead.

Because of reliance on measurement, we wound up with a variable unit of mass - the kilogram, Le Grand Kilo. We determine mass by weighing an object. A kilo weighs differently on top of a mountain than in a valley. Listen to what the International Bureau for Weights and Measures says:

"By definition, the error in the repeatability of the current definition is exactly zero...The international prototype of the kilogram seems to have lost about 50 micrograms in the last 100 years, and the reason for the loss is still unknown...It is accurate to state that any object in the universe (other than the reference metal in France) that had a mass of 1 kilogram 100 years ago and has not changed since then, now has a mass of 1.000 050 kg."

Because of this, Relativists discuss two kinds of mass, rest & relativistic, and argue over whether the mass increase is actually an increase in amount of matter or an increase in velocity.

Here, let Baez from math.ucr.edu clarify for us: "The invariant mass of a particle is independent of its velocity v, whereas relativistic mass increases with velocity... At zero speed, the relativistic mass is equal to the invariant mass... For example, when physicists quote a value for 'the mass of the electron' they mean its invariant mass... The invariant mass is therefore often called the 'rest mass'. "

In effect, what we are being told is that your mass with respect to the earth is rest mass and with respect to another planet is relativistic. In other words, your weight on earth is a measure of invariant mass and does not change since it is at rest. This means that your mass is a quantity of matter.

Wiki tells us, "The rest mass or invariant mass is an observer-independent quantity."

So conceptually we see that mass is a quantity of matter, but m in every equation from Newton to Hawking is a dynamic quantity – inertia. Relativists calculate mass of the electron after measuring its force and energy, and compare this against the kilogram.

So what changed when you moved from the valley to the mountain: quantity of matter or velocity? Neither! What has changed is the distance between you and the center of the earth and other objects near you. What has changed is static distance, not dynamic distance traveled. Change in inertia is a result of gravitational force, and force is f=ma.

The definition is circular, and what have we learned? The idea is that relativistic mass minus velocity is equal to rest mass. What is 'rest' mass without velocity if we need velocity to determine rest mass? There is no difference between rest and relativistic mass.

The only way to have rest mass is to count units of matter. There would be no ambiguity between terms. This would end the debate over mass/velocity increase. We simply count the number of units.

"The concept of 'relativistic mass' is subject to misunderstanding... it makes increase of energy of an object with velocity or momentum appear to be connected with some change in internal structure of the object." – Spacetime Physics by Taylor and Wheeler.

Relativists will tell you that mass and weight are not the same thing. Weight changes but not mass. Mass is both the amount of matter and velocity together (f=ma). They cannot separate the two, because if they did, the concept quantity of matter would destroy all of mathemagical physics.

Chapter Twenty Seven – Mass Part Two

Mass Confusion Gains Momentum

Physicists determine the mass of a particle such as the top quark by smashing it and measuring the weight (read: amount of energy needed to accelerate the quark) against a standard. The standard is Le Grand Kilo.

When an object changes locations, its weight changes in relation to the distance from the center of the earth. Consequently, so does its mass since we determine mass according to the weight we measure. The mathemagicians will tell you that its velocity or energy changed because they don't know how to quantify matter and so they cannot tell you how many units of matter an object has.

If we were counting units of mass, we could determine quite easily if mass increased by the additional units. If mass did not increase, then we would understand what football players have understood for decades, that more pressure is exerted at faster speeds. Instead, Relativists break up inertia into two additional qualifiers, invariant and relativistic mass, confusing themselves and others.

So what is Special Relativity's equation predicting? Is an object's mass actually increasing with speed, velocity, or quantity? With math magic anyone can predict any physical theory they want and they will all be right, but they are giving qualitative meanings to quantitative equations. One can do anything they want in thought space that cannot be done in thing space.

There are actually four physical interpretations for mass increase in Special Relativity. The first three we covered in Mass Confusion Part One.

The traveler:

a) at near the speed of light experiences no change in inertia

b) has greater inertia because of velocity or location (f=ma)

c) gains more substance

d) experiences an increase in mass due to an increase in energy

"[Einstein] did not introduce the notion that the mass of a body increases with velocity — just that it increases with energy content." – math.ucr.edu

Len Bugel from Teaching Relativity in High School says, "Another interesting consequence is that as you continue to add energy to a particle traveling near light speed, the particle stores this energy as an increase of mass rather than an increase in velocity. Protons entering Fermilab's Tevatron with 150 billion electron volts (GeV) of energy are already traveling at 0.99998 times the speed of light and have a mass of about 150 times their rest mass. When they leave, the energy and mass have increased by a factor of more than 6, while their speed has barely changed."

What is this mysterious energy, and why does it increase? I tell you more about it in Part Three.

"While photons have no mass, they do possess momentum." - NASA

"Light is composed of photons, so we could ask if the photon has mass. The answer is then definitely 'no' - the photon is a massless particle. According to theory, it has energy and momentum, but no mass... light carries momentum and will exert pressure on a surface. This is not evidence that it has mass since momentum can exist without mass." – physicsforums.com

If a photon had any mass, the Lorentz-Fitzgerald equations dealing with length contraction would flounder. However, the photoelectric effect had to be explained, so mathemagicians used momentum...hence relativistic mass.

The Relativists manipulate Einstein's equation by adding momentum and connecting that mysterious concept: energy. Here is the equation representing the energy of a particle where it resides (its frame of reference) from Noble Prize.org

$E\ 2 = (m_{rest} * c\ 2)\ 2 + (p * c)\ 2$

rest mass + relativistic mass

E is energy, mrest the rest mass of the particle, p is momentum, and c is the velocity of light

You see that there are two types of mass. Rest on the left and relativistic on the right (p = mrel v * c). Setting the first term (mrest) to zero we get 'mass without mass' (sort of like Bruce Lee's 'Fighting without Fighting' from Enter the Dragon) but the particle has motion (relativistic mass). In other words, we are talking about the photon. What they are really saying is that rest mass is mass a la Newton, and relativistic mass is really energy. Our little buddy the photon has no mass, but lots of that magical substance, energy.

Energy is equivalent to mass.

What's energy? Relativity folks think that it exists, but that it does not have shape! Now would be a good time to review what it means to exist:

Exist: object with location.

Location: static distance between all objects

Object: that which has shape

Energy is not a thing, substance, or material object. Energy is not a noun, it is a verb. Mathematics is a language that describes dynamic concepts.

By setting (mrest) to zero on the left side of that equation, the term (mrest * c2) 2 (p*c)2:

$$E2 = (mrest\ c\ 2)\ 2 + (\ p\ c\ )2$$

$$= 0 + (\ p\ c\ )2$$

$$E = (p)\ c$$

$$= (mrel * velocity\ of\ object) * c$$

We are left with E = momentum * velocity of c. Momentum is mass * velocity. So after all that, we still end up with mass. At this point the relativist will tell you, "Relativistic mass is really energy."

What he or she is trying to get you to swallow is that you can have velocity without any amount of matter. Since our friend the photon has no rest mass, he can never be at rest can he? If he stops moving, he no longer exists. Without rest mass there can be no relativistic mass. Photon has no rest mass, but he has infinite relativistic mass. He is pure energy! See how good our magic word energy is? Rest and relativistic mass are really at opposite ends of the scale of mass going from 0 to c.

"The relativistic mass is just equal to the rest mass if the particle is not moving." – from a lecture at mtsu.edu

Here's another scenario: velocity without velocity. If we set $(p\ c)2$ to zero (zero velocity) we have no momentum and what is left? Mass.

$$E2 = (mrest\ c\ 2)\ 2 + (p\ c)2$$

$$= (mrest\ c2\ )2 + 0$$

or $E = mrest\ c2\ [E = mc2]$

Mass without motion is a contradiction because mass is a dynamic parameter. So whether rest mass or momentum is set to zero, we will end up with mass. So what did we learn? We learned that mass increase is due to mass increase - sure Einstein. If you ask the Relativist, "What if we set the mass to zero, wouldn't energy be zero?" He will tell you that energy is actually greater than zero. When you ask how this is possible since energy and mass are equivalent, he will tell you that it was actually momentum that increased. If you tell him you thought mass was necessary for momentum, he will tell you a photon doesn't have mass, it has energy (relativistic mass).

In Special Relativity, energy and momentum doesn't tell us what actually increased because of the use of synonymous terms. Mass is energy, but energy can increase without mass because all we need is momentum - but momentum is mass times velocity. A photon has no mass, but it has energy, and by increasing its momentum we increase its energy.

Kewl mathemagical slight of hand, no? By using the elusive terms energy and momentum, we can have mass without mass or velocity without velocity. Energy without mass (E= p c) and energy without mass (E = m c 2). Got it? Good, now don't you feel silly for asking?

Let's cover one last thing before moving on to the magic of energy. The photon is a particle without mass or shape, but it moves. Sure. How does the zero dimensional photon magically transform into the three dimensions of reality?

"In quantum theory, particles can be created out of energy in the form of particle/antiparticle pairs." Brief History of Time, Stephen Hawking (p. 129)

No one can tell you how a zero dimensional particle suddenly acquires length, width, and height. If you think that's funny, momentum is a real joke. Momentum, which is the motion of an object, has been turned into a noun. You don't believe me?

"ħ may be said to be the 'quantum of angular momentum'." - Wiki

Therefore, mathemagicians think that momentum has shape. Quantum of momentum...units of momentum...momentum moves!

Not tired of laughing? Good, wait until you read the next chapter on energy Magic!

Chapter Twenty Eight – Mass Part Three

Here, go get me some energy please!

Energy According to Relativity

The reason energy is such a mathemagical term is because it remains undefined or is defined ambiguously.

"It is important to realize that in physics today, we have no knowledge of what energy is."(Ch. 4-1) – Richard Feynman from Feynman Lectures Physics Vols 111

This all purpose word really comes in handy for the mathemagician and can be used in many of his or her incantations.

"It was generally construed that all changes can in fact be explained through some sort of energy. Soon the idea that energy could be stored in objects took its roots in scientific thought, and the concept of energy came to embrace the idea of the potential for change as well as change itself." - Wiki

Never has there been a word that "explains" so much yet means so little. If there is a phenomena, concept, or idea, and ordinary words just don't do them justice, invoke the magical term of energy. No one can dispute that word. Everything is energy: radio waves, X-rays, gamma rays, Manta Rays, gravity, heat, field, space, you name it! We have potential energy, kinetic energy, nuclear energy, thermal energy, magnetic energy, electrical energy, and gravitational energy.

What does it have to do with Special Relativity's prediction of mass increase with speed? Let's see if we can find out.

Is energy an object?

"By assuming that light actually consisted of discrete energy packets, Einstein proposed a linear relationship between the maximum energy of electrons ejected from a surface and the frequency of the incident light." – aps physics.org

Note the use of discrete energy packets, which clearly shows that the mathemagician sees energy as having shape, that energy is an object by definition - Object: that which has shape.

Note also the use of language here which indicates that energy can be carried like the groom carries his bride across the threshold or like one carries a bucket of water from a well.

"[Faraday] realized that electric and magnetic fields are not only fields of force which dictate the motion of particles, but also have an independent physical reality because they carry energy." - wiki

The Laws of Thermodynamics say that energy cannot be created or destroyed, Obviously, if energy can be carried and cannot be created or destroyed, then it must be a physical object. Of course it may be argued that physicists are using euphemisms. So what do they really mean? Why not just tell us what energy really is?

"Energy is so fundamental that it is not easily defined in terms of anything more fundamental... For a general audience, rather than worrying about the details of a formal definition, it is far easier and far more useful to understand what energy does in various situations. This is called the 'energy is as energy does' school of thought." – also from the Wiki link on energy above.

Mathemagicians rely on the operational definition of the word energy, yet use it as though it were a physical object. Sorry, they can't use this word in physics until they can define the word energy consistently and without ambiguity. If energy is a concept, it cannot move or be carried. If it is an object, it must have shape.

"The energy of the photon refers to a particle" (p. 91) - About Time: Einstein's Unfinished Revolution

If energy is a discrete particle, then it obviously has shape, a border. If so, we can draw a picture of it when it is not moving. We can imagine it moving (two or more locations). We all use the term energy as a concept when we say, "I'm tired, and I have no energy."

$E = mc^2$

Clearly, energy, mass, and the speed of light are not objects. They are concepts. Physics is the study of physical objects.

Look at the definitions for energy in a dictionary. Wolram science dictionary says, energy is a concept:

"Energy is an abstract quantity of extreme usefulness in physics because it is defined in such a way that the total energy of any closed physical system is always constant (conservation of energy). It is impossible to overstate the importance of this concept in all branches of physics from elementary mechanics to general relativity. Energy is measured in units of mass times velocity squared."

We probably all remember being told in an elementary school science class that energy is the ability to do work. Does ability have shape? An ability like energy is not what something is, it is what something does. A fish (noun) has the ability to swim (verb). Can you draw a picture of "a swim"? Does swim have shape?

What about the adverb 'capacity' that is constantly being thrown around with the word energy? An aquarium may have the capacity to hold 20 gallons of water, but does capacity itself have shape, or is it the aquarium that has shape? How can we transfer energy like transferring the capacity of the aquarium to another aquarium? We can't, but we can transfer the 20 gal. of water to another 20 gal. aquarium with a bucket.

In the magical world of relativism, energy is an ability, an object, a capacity, and it is also motion. In reality, energy cannot be all of these things at the same time. Energy explains everything, yet means nothing!

Actually, according to Hawking in A Brief History of Time, energy is nothing:

"The inflation was also a good thing in that it produced all the contents of the universe quite literally out of nothing. When the universe was a single point, like the North Pole, it contained nothing... relativity and quantum mechanics allow matter to be created out of energy in the form of particle/antiparticle pairs." (p. 97)

"In quantum theory, particles can be created out of energy in the form of particle/anti-particle pairs. But that just raises the question of where the energy came from. The answer is that the total energy of the universe is exactly zero. The matter in the universe is made out of positive energy. However, the matter is all attracting itself by gravity... the gravitational field has negative energy... this negative gravitational energy exactly cancels the positive energy represented by the matter. So the total energy of the universe is zero." (p. 129)

The magical zero, and the even more powerful word 'infinity', always come in handy to the relativist. Before the universe was nothing (0), and the energy appeared out of nowhere (infinity).

In conclusion, when a relativist tells you that there is an increase of mass with an increase of speed and an increase of energy, he has explained nothing at all. As I said in Mass Confusion Part One:

What is energy? Relativists don't know what that is. Yet they will try to convince you that the so-called mass increase that they don't understand is really energy increase that they also do not understand.

Chapter Twenty Nine – Big Bang Theory

The Big Bang Theory by Bare Naked Ladies

Our whole universe was in a hot dense state,
Then nearly fourteen billion years ago expansion started.
Wait... The Earth began to cool,
The autotrophs began to drool,
Neanderthals developed tools,
We built a wall (we built the pyramids),
Math, science, history, unraveling the mysteries,
That all started with the big bang!

Astronomer Fred Hoyle ridiculed Lemaitre's theory by coining the term Big Bang. Einstein took issue with Lemaitre's math and told him that his physics was "abominable", and his conclusions were not justifiable. If you think this should have been the end to the theory since it was supposedly founded on Einstein's theory of General Relativity, then you are thinking rationally.

Of course theorists, mathematical physicists, and cosmologists want you to believe that the universe doesn't make sense in an attempt to get you to abandon your common sense and rational thinking. Don't fall for it! One thing that sets mankind apart from all the rest of God's creatures is our ability to think critically and rationally. Just because mainstream science wants to abandon rationality, does not mean that you must. Stand up to these bullies that ridicule your beliefs. Use your common sense and their logic along with science to destroy the religion of the Big Bang Theory (BBT)!

Oddly, even though mainstream science depends on observation, testing, and prediction, the BBT fails major tests and predictions. If the universe was expanding, then we should expect to see galaxies traveling away from a central point at the same speed. There are hundreds of galaxies that are blue-shifted, traveling towards our galaxy. There are colliding galaxies, galaxies orbiting each other, and galaxies that are 'clumping' together.

There are many other problems, such as the fact that there is not enough mass and density in the universe.

When faced with these and other observations indicating failure of the predictive power of BBT, instead of scraping the theory and starting over, cosmologists came up with the invisible, undetectable Dark Energy and nonsense such as multiple universes! No one quantitative prediction has ever been observed. Cosmologists invent more and more parameters which they constantly adjust in an attempt to make their theory conform to their observations. This is not how science should work. When a theory fails, then the theory should be tossed out, and the theorist should start over.

Of course the Big Bang theory doesn't answer many questions, like what happened before the explosion, where the singularity came from, why it happened when it did, and how the theory can be reconciled with other accepted scientific theories such as the one that says matter and energy cannot be created or destroyed!

There are many difficulties with the BBT, such as the cosmic microwave background being more an indication of heat from starlight than the aftermath of some giant explosion, there are too many voids in space to account for the amount of time allotted to them, it is unlikely that average luminosity of quasars decrease in the same way so that all red shifts have the same average brightness, globular clusters are older than the universe, ratio of density to matter may indicate the universe should have already collapsed or dissipated, etc., etc.

There are simply far too many failed predictions for the current model to be accepted over other competing models, such as; Quasi-state, Variable Mass Cosmology, Meta Model, and Plasma Cosmology.

Logically, the BBT is full of space and time fallacies:

Fallacy of self reference

Fallacy of reifying abstractions

Fallacy of confusion between the properties of a set and properties of the members of a set.

But wait! We don't need to use any of the above. We need not use formal logic or math or mainstream science to destroy the arguments coming from the proponents of the failed Big Bang Theory. We can end the argument once and for all conceptually.

Space cannot expand, as it is nothing. 'Nothing' cannot expand, as it does not exist. What would space expand into, more space? And if space is something, then it would be a large block of matter where motion would be impossible. Still we are left wondering where that block of matter 'sits.' In a bigger block of matter?

Equally irrational is the idea that time began with the Big Bang. Time is not a thing, it is a concept created by man, or as Einstein said, "Time is what clocks do." Yet the irrational theorist will tell you that both space and time began at the Big Bang.

At the heart of it is what cosmologists call the singularity. In The Universe In A Nutshell, Hawking explains that he and Roger Penrose came up with the singularity to avoid the problems of time and space being infinite (Something interesting to check out is Kant's Antinomy of Pure Reason.) The singularity hinges on the ridiculous assumptions of Quantum Mechanics and Quantum Gravity, where "the quantities used to measure gravitational fields are infinite."

Infinities are irrational and therefore impossible as pointed out in the chapter "Infinite Space Revealed."

Here is the definition of singularity from the Singularity Symposium: "A point or region in spacetime in which gravitational forces cause matter to have an infinite density; associated with Black Holes."

Relativity combines the three dimensions of reality (height, length, and width) with time into a fourth dimension and calls it spacetime. Of course any rational person can understand that in reality there are only three dimensions. Space is void, and time is a concept. Can anyone imagine the fourth dimension called spacetime?

You can't draw it:

"The problem with visualization becomes even more acute when we try to graph the data. You can make a drawing of a three-dimensional object on a flat piece of paper, and you can even make a model to represent three-space, like a relief map that shows the elevation of the ground at every point over a given area. But when you try to draw a picture of four-dimensional space it is impossible."

You can't visualize it. According to mathforums.org:

The only reason people get confused about it is because they cannot visualize it. If I tell you the length, width and height of an object, you can get an idea of what it looks like, perhaps a cube or a long slender rod or anything in between. But if I also tell you what temperature something is or how much it weighs, what does that look like?

And according to Stephen Hawking, you can't even imagine it:

"An event is something that happens at a particular point in space and at a particular time. So one can specify it by four numbers or coordinates...one can use any three well-defined coordinates and any measure of time. In relativity, there is no real distinction between the space and time coordinates...It is often helpful to think of the four coordinates of an event as specifying its position in a four-dimensional space called spacetime. It is impossible to imagine a four- dimensional space." (A Brief History In Time, Stephen Hawking, p. 2x)

A singularity with 0-dimensions is impossible. Neither are 1-dimensional nor 2- dimensional singularities possible as postulated by Hawking and Penrose. Now can anyone explain how length, width, and height can spring forth from zero dimensions? No, no one can.

No one can draw spacetime, no one can visualize spacetime, and no one can imagine spacetime. Neither can anyone imagine zero-dimensions or four dimensions. Why? It is irrational! There are only the three dimensions of reality, that's why!

Which came first the Cosmos or the Cosmic Egg?

Lemaitre's BBT was originally called the Hypothesis of the Primeval Atom, otherwise known as the cosmic egg. BBT proposes that all matter, energy, time, and space were in a hot, dense state (the size of a dimpled pea as Hawking has described it). So where did this cosmic egg come from? Proponents of BBT claim that it is not creatio ex nihilo, created from nothing, as seems obvious. So then where? Their answer is they can't know what happened before the BB, but some guess our universe came from another universe, and that one came from another universe which came from another universe, and so on ad infinitum. They want you to believe there is an irrational, infinite regress of universes or an infinite nesting of universes.

"A quantum fluctuation is the temporary appearance of energetic particles out of empty space, as allowed by the Uncertainty Principle." - Wikipedia

Right! Virtual particles and entire universes popping out of no-where and no-when.

Or, as physicist Lawrence Krauss, author of A Universe From Nothing has said, "So space—indeed, there was nothing in the conventional sense that there was no space, no time, and no universe. It's perfectly plausible that a universe can be created where there was no space before. In fact, again, in quantum gravity, it's not only plausible; it's required! It's required that you cannot have that event not happen somewhere. But the laws are there."

BUT...space is not really nothing:

"...it would be disingenuous to suggest that empty space endowed with energy, which drives inflation, is really nothing." (Lawrence M. Krauss, A Universe From Nothing, p. 152).

No kidding. And yet, space is nothing:

"As I have defined it thus far, the relevant 'nothing' from which our observed 'something' arises is 'empty space.'" (Lawrence M. Krauss, A Universe From Nothing, p. 161).

"...a quantum theory of gravity allows for the creation, albeit perhaps momentarily, of space itself where none existed before." (Lawrence M. Krauss, A Universe From Nothing, pp. 163-164).

Nothing means something different than it used to:

'Why is there something rather than nothing?' must be understood in the context of a cosmos where the meaning of these words is not what it once was, and the very distinction between something and nothing has begun to disappear. . ." (Lawrence M. Krauss, A Universe From Nothing, pp. 182-183).

Nothing is invisible dark matter...:

"We have discovered that 99 percent of the universe is actually invisible to us, comprising dark matter that is most likely some new form of elementary particle, and even more dark energy, whose origin remains a complete mystery at the present time. . .Maybe literally, as well as metaphorically, we are making much ado about nothing. At least we may be making too much of the nothing that dominates our universe!" (Lawrence M. Krauss, A Universe From Nothing, p. 138)

...whatever that is...

The follow gems of wisdom also from A Universe From Nothing.

We understand nothing about nothing:

"After all, since we have no idea what the dark energy permeating empty space is, we also therefore cannot be certain that it will behave like Einstein's cosmological constant and remain constant."

"Cosmology has produced one totally mysterious quantity: the energy of empty space, about which we understand virtually nothing."

"The origin and nature of dark energy is without a doubt the biggest mystery in fundamental physics today. We have no deep understanding of how it originates and why it takes the value it has."

There are different kinds of nothing: "First, I want to be clear about what kind of 'nothing' I am discussing at the moment. This is the simplest version of nothing, namely empty space. For the moment, I will assume space exists, with nothing at all in it, and that the laws of physics also exist. Once again, I realize that in the revised versions of nothingness that those who wish to continually redefine the word so that no scientific definition is practical, this version of nothing doesn't cut the mustard. However, I suspect that, at the times of Plato and Aquinas, when they pondered why there was something rather than nothing, empty space with nothing in it was probably a good approximation of what they were thinking about." (Lawrence M. Krauss, A Universe From Nothing, p. 149).

Yep, space 'exists' with 'nothing' in it!

There you have it folks directly from the expert on nothing!

Chapter Thirty – Black Holes

Do They Exist?

Well, that depends on two things: the definition of black hole and the definition of exist. Before defining these terms, let's take a brief look at the varying opinions. I Googled, "Do black holes exist?" and randomly selected these top links. They probably have, might not have, apparently have, cannot have, and have been discovered.

MacDonald Observatory: "Do black holes exist? Probably. Astronomers have discovered quite a few objects that can only be explained as black holes. These objects are dark, so we cannot see them, but they exert a powerful influence on stars, gas, and even space around them. These objects are so dark, dense, and heavy that they must be either black holes or something even more exotic."

NewScientist: "Black holes might not exist - or at least not as scientists have imagined, cloaked by an impenetrable 'event horizon'. A controversial new calculation could abolish the horizon and so solve a troubling paradox in physics."

PBS - Stephen Hawking's Universe: "A supermassive black hole with 2 billion times the mass of the sun apparently lurks in the nearby giant galaxy M87. ...stars move about so quickly that they must be caught in the grips of a massive object. By calculating the size and mass of these objects, the only conclusion seems to be that the center of these galaxies harbor supermassive black holes."

Nature: "Black holes are staples of science fiction and many think astronomers have observed them indirectly. But according to a physicist at the Lawrence Livermore National Laboratory in California, these awesome breaches in spacetime do not and indeed cannot exist."

Harvard: "They say that truth is stranger than fiction, and it turns out that nature is stranger than science fiction. More than a dozen black holes have already been discovered in our Milky Way galaxy - out of more than a million black holes estimated to exist there."

Why are there uncertain or conflicting answers coming out of McDonald Observatory, NewScientist, PBS, Nature, and Harvard? Why are we left with more questions than we started out with? Physicists have been theorizing and astronomers have been looking for black holes for over a hundred years.

What is a black hole? There are many varying definitions:

Free Dictionary: "An area of spacetime with a gravitational field so intense that its escape velocity is equal to or exceeds the speed of light."

Dictionary.com: "A theoretical massive object, formed at the beginning of the universe or by the gravitational collapse of a star exploding as a supernova, whose gravitational field is so intense that no electromagnetic radiation can escape."

Merriam Webster: "A celestial object that has a gravitational field so strong that light cannot escape it and that is believed to be created especially in the collapse of a very massive star."

So what is it: an area of spacetime, a theoretical massive object, or a celestial object? We are told that a black hole is an area or a region of space with so much mass that nothing, including light, can escape its gravitational pull.

To understand what cosmologists are trying to tell us we need to look at object, spacetime, mass, and singularity. There is confusion about the black hole, and it is because of the contradictory, ambiguous terms being used. Area is a concept, and object is that which has shape, mass is a property of matter, and time is a concept relating motion of objects and memory. At the center of the black hole is a singularity. What is that?

That is an important question. The deeper we look, the more questions we have, in a never ending search for what we thought was a simple question looking for a simple answer. Of course, we are told that there are no simple answers. The universe is a strange place and does not conform to our expectations of being logical and rational. Don't buy it!

Black holes depend on singularities, and singularities depend on relativity. We took a look at relativity in the chapter "E=MC Squared Away."

In physics it is crucial to understand what that little word 'exist' means, and Einstein rightly points out that whether or not there is an observer has no bearing on what exists. What does it mean to exist? This is the most basic question of science in general and physics in particular. One would rationally think that scientists could tell us. However, astonishingly they cannot! Physics, which is the study of what is physically present, reality, doesn't have a clue!

Herein lies the reason for varying opinions on whether or not black holes exist. This is why we get ambiguous or conflicting definitions from mainstream science. We took a more in depth look at the relationship between objects and concepts in the chapter "The REAL Scientific Method."

Cosmologists and theoretical physicists live in the make believe world of thought space where abstract concepts and imaginary numbers bump into each other instead of living in thing space where real objects are located.

Black holes are impossible because singularities are impossible. Zero dimensional point particles are mathematical constructs and a figment of the imagination.

We discussed singularity and spacetime which is at the heart of General Relativity and the Big Bang Theory in the chapter "How Theists Can Destroy Big Bang Creation."

Relativist equations predict that the magical, mysterious black hole must exist. However, their equations are abstract mathematical concepts which supposedly can suck real objects, including the bathroom sink and light, along with it. How do physicists reconcile a region of space with an object?

"A black hole is a simple object that has only a 'center' and a 'surface.' " (Universe, William Kaufmann, p. 469)

"[Penrose] showed that a star collapsing under its own gravity is trapped in a region whose surface eventually shrinks to zero size. And since the surface of the region shrinks to zero, so too must its volume. All the matter in the star will be compressed into a region of zero volume…In other words, one has a singularity contained within a region of spacetime known as a black hole." (Stephen Hawking, A Brief History Of Time, p. 49)

A region with zero size? Right.

Some more black hole absurdities:

Black hole theory says that a black hole has an escape velocity of the speed of light in a vacuum, yet we cannot see a black hole. We are told light cannot escape from it!

Since it takes an infinite amount of observer time for a body to reach the event horizon, and since no observer can possibly be around for an infinite amount of time to make the observation, the event horizon is meaningless.

Black holes contain an infinitely dense point-mass singularity, yet Special Relativity forbids infinite density as does rationality.

So what is a black hole? It is an abstract mathematical concept with no corollary in reality.

What does it mean to exist? To be physically present, that is to exist, is to be an object with a location.

Do black holes exist? Obviously not!

The Black Hole on trial: The prosecution sums up its case.

Your Honor, Ladies and Gentleman of the jury, people of the court, the prosecution alleges that Black Hole is a fraud concocted by priests, promoted by the Rock Stars of physics, and perniciously perpetuated against science and society.

The prosecution has shown that Black Hole fails at every level. Singularities fail conceptually, as zero volume means no length, width, or height- an irrational proposal with no corollary in reality. Infinite density fails even the most basic math as one cannot divide by zero as required by the simple math formula of density, mass, and volume, and infinities cannot exist in reality regardless of what mathematicians do with 'higher math.' Black Hole also fails at the higher level of Newtonian and Einsteinian theory and corresponding maths. Black Hole can only be seen due to so-called gravitational lensing which is circumstantial at best, and then only as an ad hoc presentation in lieu of observation of an event horizon or Black Hole itself.

Let's review the case.

Defense Exhibits:

Exhibit A: They can be predicted by theory.

Exhibit B: They can be indirectly observed.

Exhibit C: There are no alternatives.

The defense alleges that Black Hole is predicted by theory. Their Exhibit A states:

"Karl Schwarzschild created the first modern resolution of relativity that would characterize a black hole in 1916, and later work from many physicists showed black holes are a standard prediction of Einstein's theory of general relativity."

This is clearly not so, as attested to by Karl Schwarzschild himself in Mr. Schwarzschild's paper entitled 'On the Gravitational Field of A Mass Point According to Einstein's Theory', and confirmed by Leonard Abrams' paper 'Black Holes: The Legacy of Error.'

Einstein's Special Theory of Relativity clearly prohibits infinite densities. Karl Schwarzschild attests to this fact, proving it in his paper on point mass. Not only this, Einstein denied the possibility of black holes multiple times before his death in 1955. Both Einstein Relativity and Schwarzschild's solution theories forbid infinite densities.

So does common sense and rationality. Although math can postulate infinite densities in abstraction, reality is having none of it. While Hilbert can build hotels in thought space with an infinite number of rooms, the world of reality does not comply. Though Zeno in his Dichotomy Paradox can halve a distance infinitely in abstract mathematical equation space, one cannot walk half way to a brick wall indefinitely and certainly cannot halve their distance an infinite number of times. All who have tried end up smacking their foreheads on the wall.

Furthermore, later work in the 40's conveniently ignored relativity and erroneously posited an "infinity of spacetimes differing as to the limiting acceleration of a radially approaching test particle." In other words, Hilbert substituted a variable with a scalar invariant transforming the coordinate location of a point mass. Because of the error, the point $r=0$ becomes a two-sphere invalidating Hilbert's assumption.

Owing to the extreme difficulty of calculus and other maths involved, our expert witnesses have reduced the equations to simpler language for the layperson. There are no known solutions for Einstein's field equations for two or more bodies, and yet proponents of Black Hole allege multiple masses interacting with each other and with matter. The principle of superposition applies to Newtonian masses but not to General Relativity, so Newton's escape velocity cannot be used in an expression relating to a universe containing only one mass. Einstein's theory, as Schwarzschild shows, pertains to one mass. In other words, Newton's theory contains two masses and superposition. However, $r=0$ contains no bodies, and so therefore cannot accommodate superposition.

The defense exhibit A, predicted by theory, fails on multiple counts. The defense states that Black Hole can be indirectly observed. The so-called evidence cited is that because light can escape Black Hole's event horizon, one must look for gravitational lensing. Of course as the prosecution has already shown, Newton's escape velocity has no relevance in a universe of only one mass as required by Einstein and Schwarzschild's theories.

The idea of an escape velocity of light also fails logically as previously stated. If the escape velocity from an event horizon is the speed of light, then light can escape an event horizon.

Although eyewitness accounts can be effective in swaying a juror unknowledgeable in these matters, it is well known that eyewitnesses are among the least reliable sources.

This is worse than that because we have witnesses claiming they saw where alleged black holes are circumstantially. Furthermore, no one has ever found or photographed an event horizon or a black hole, instead they have provided artist renditions of a black hole or telescopic images of unidentified objects ad hoc in an attempt to save their failed theories.

Therefore the defense exhibit B, they can be indirectly observed, fails on multiple counts.

The defense presents Exhibit C: there are no alternatives. We are told "Very few physicists would tell you there are no black holes in the universe." The defense admits there are alternative theories to Black Hole when council states, "Certain interpretations of supersymmetry and some extensions of the standard model allow for alternatives to black holes." And yet we are reminded, "But few physicists support the theories of possible replacements." At one time very few scientists would tell you that the earth was round, but that does not make the earth flat.

The burden of proof falls on the one making a claim, not on another person who forms a different conclusion. Since the prosecution has shown the premises to be false, it follows that the conclusion is false. Furthermore it is a fallacy, a non sequitur, it does not follow that because we don't have an alternative, a theory - let alone an invalid theory - is correct. Regardless, in science when the hypothesis or theory fails, one erases the whiteboard and starts over.

Gravitational lensing is supposedly caused by dark matter, itself an unproven and irrational proposition at the hypothesis level. It's not even a theory yet!

Lastly, there are multiple theories of gravity. Therefore gravitational lensing is not a 'settled' issue to begin with. Not only that, but no satellite including the gravitational wave observatory LIGO has ever detected a gravitational wave.

Additionally, Gravity Probe B did not confirm frame dragging, distortion of spacetime around a large body. The folks there used a hypothetical model to show why they didn't find anything, and then years later altered the data to claim they proved frame dragging.

Therefore the defense Exhibit C fails on multiple levels.

Exhibits for the prosecution:

Exhibit A: Fails conceptually.

Exhibit B: Fails at basic math.

Exhibit C: Fails at higher math.

Now for the prosecution's summary. Not only does the council for the defense fail to prove Black Hole is possible, the prosecution has presented evidence that Black Hole is founded on failed hypotheses and is in opposition to accepted theories.

Black Hole fails on many levels. It fails at the conceptual level: it is illogical and irrational. Infinities are impossible and so is zero volume. Black Hole fails at simple math: one cannot divide by zero. Black Hole fails at higher math: it is in violation of Einstein's relativity, Schwarchild's solution and classical Newtonian physics, and both superposition and escape velocity equations.

All this is confirmed by Einstein and Schrawzschild and born out by other physicists which have provided expert testimony in their stead.

In conclusion, the defense, having not made a case, and the prosecution having shown beyond a reasonable doubt that Black Hole is impossible; the jury must find the Defendant Black Hole guilty as charged, and must find that Black Hole is non-existent!

Chapter Thirty One - The Nature of Light

The Phiz Whiz Is Not the Brightest Bulb on the Tree

Is light an object? Is it phenomena? Is light a particle, a wave, both, or neither? Let's a take a quick look at the history of light science, and see how far it's come in the last 400 years or so.

Light behavior is strange and illusive. Two flashlight beams cross each other, yet a child can stop the light with his hands when he makes shadow puppets. For hundreds of years scientists and philosophers have pondered and mused over it.

We are told in physics class about the dual nature of light, that light has properties of both a particle and a wave. Light is a wave until we observe it, then it becomes a particle!

Heron of Alexandria, in the first century said, "Everybody thinks that light travels in straight lines. Well, all fast-moving objects travel in straight lines." Early on, the Greeks thought that vision left the eyes and traveled to the object.

Ptolemy noticed that there is a ratio between angles of incident and refracted light, but this only held true for some angles. Willebrord Snellius' law says that the ratio is constant for specified materials such as glass, air, or oleic acid. Today, scientists and philosophers are still studying light behavior, with no real understanding of what light actually is.

Aristotle and Lucretius thought that light particles came from the sun to the eyes. About a thousand years later Alhazen explained that light came from objects to the eyes.

Later...Descartes revived Alhazen's refraction laws that relates angle of incident and refracted light to velocity (now referred to as Snell's Law). Ever since that time, mathemagicians calculate rather than measure the velocity of light.

The seventeenth century researchers thought that light traveled instantaneously, or nearly instantaneously, from here to there. Roemer explained that there was a delay in the eclipse of Jupiter's moon, Io, between May to October, because there was a greater distance between Io and the earth, showing that light speed is finite.

Hooke and Huygens thought that light travels as a wave producing secondary wave-fronts. Newton thought that a ray of light was comprised of corpuscles and travels in a straight path.

Newton's theory predicted that light would travel faster through a denser medium because sound travels faster through water than through air. Huygens disagreed with the particle theory of Newton because light beams crossed through each other without apparently dislodging particles of light. He wondered why water or glass reflected some light and allowed some to pass through.

Huygens was a wave man because, to him, particles didn't explain diffraction, interference, or color. Newton was a particle man because spreading wave-fronts didn't jive with the rectilinear path of light. How does light bend around objects like water and sound waves?

Nineteenth century experiments continued to indicate that light could not be a particle, but the mathemagicians couldn't let go of the notion, and held the particle near and dear to them. Their equations relating mass as a point particle just wouldn't allow them to abandon their love affair with the particle theory of light. She would forever be theirs.

Young's slit experiment all but demolished Newton's corpuscular theory of light in 1801. Fresnel's wave theory of light explained interference and he entered it into a contest in an attempt to win the Grand Prize of the French Academy of Science. Poisson, a particle light man, was a member of the academy and tried to disprove Fresnel's theory with the "Poison Spot." The head of his committee, future Prime Minister of France, Arago, was able to convince most of contemporaries of the wave nature of light with an experiment where he observed Fresnel's predicted spot. Mathemagicians, to this day, love to observe and to use their position, authority, or committee, to convince others. AragoBarthelonius discovered polarization and explained it with transverse waves which contradicted Newton's longitudinal waves. Fresnel's theory of polarization stomped on the particle theory because it demonstrated that light apparently consisted of two transverse waves perpendicular to each other.

The wave nature of light theory was further strengthened by Oersted, Faraday, and Maxwell, which indicated magnetic fields running perpendicular to electric fields. Maxwell proposed that the visible range of the electromagnetic spectrum is what we called light.

But the particle theory of light, like a jilted lover complained yet returned for more. Max Plank's proposed discrete packages called light quanta, and Einstein used that to explain the photoelectric effect (shooting an intense beam at a metal surface dislodging electrons from atoms on the surface of the metal). Wave theorists thought inducing current flow would result in a gradual transfer of energy and electrons dislodged would be proportional to the intensity of the beam. Einstein said that the number of electrons dislodged were proportionally related to the frequency of the light striking the surface. So here we have a stream of electrons, and a frequency dependent photoelectric effect; in essence both particle and wave phenomena.

Enter Rutherford's planetary model of the atom and Thomson's plum pudding model of the atom. Thomson's mass-less cloud of charges won out over Rutherford's negative charges orbiting a positive nucleus for a time because Newtonian mechanics predicted that electrons would loose energy and quickly collide with the nucleus.

Bhor comes along in 1913 and postulates that when an electron falls to a lower orbit around the nucleus it looses energy and when it moves back to a higher orbit it gains energy thereby eliminating Thomson's concern of instability. Now Rutherford's model was more appealing to the scientific community. However, Harvard's Tower Experiment predicts that light leaving the earth looses energy (because of gravity) and gains it when approaching the earth. This is the opposite of what Bhor's atom is supposed to be doing.

Milikan's oil drop experiment, Wilson's cloud chamber, and Compton's X-Rays further strengthened the particle view of the sub atomic world. But, the phiz whiz's love affair with both the wave and the particle was not over. Compton's X-rays, like Einstein's explanation of the photoelectric effect, depend on wavelength.

DeBroglie "discovered' that the electron also had a wave like nature:

"If we begin to think of electrons as waves, we'll have to change our whole concept of what an "orbit" is. Instead of having a little particle whizzing around the nucleus in a circular path, we'd have a wave sort of strung out around the whole circle. Now, the only way such a wave could exist is if a whole number of its wavelengths fit exactly around the circle. If the circumference is exactly as long as two wavelengths, say, or three or four or five, that's great, but two and a half won't cut it." – University of Colorado at Boulder

Schrodinger's equation allowed the phiz whiz the ability to determine the probability of where one may find an electron in its orbit. On the other hand, Pauli's Exclusion Principle, stating that electrons with the same spin can not occupy the same space, contradicted DeBroglie's matter waves.

Not to worry, Born and Heisenberg figured out a way for the theoretical physicist to stay married to the particle and still keep his mistress, the wave, on the side.

It seems the electron is neither a particle nor a wave, but a cloud enveloping the nucleus. According to Quantum Mechanics, it's not really a cloud either, the wave function is an ABILITY to locate a particle:

"At the instant of time when the position is determined, that is, at the instant when the photon is scattered by the electron, the electron undergoes a discontinuous change in momentum. This change is the greater the smaller the wavelength of the light employed, i.e., the more exact the determination of the position. At the instant at which the position of the electron is known, its momentum therefore can be known only up to magnitudes which correspond to that discontinuous change; thus, the more precisely the position is determined, the less precisely the momentum is known, and vice versa." (Heisenberg, 1927, p. 174-5).

QM's Complimentarity states that light can behave as both a wave and a particle, just not at the same time (in the same experiment).

So after hundreds of years the mathemagicians are still talking about how light behaves, not what light is. They bypass the hypothesis stage by not illustrating the object, instead they are describing (not explaining) the phenomena. The theorist uses the functional definition of light instead of explaining what light is. Uncertainty and Complimentarity identify particle with position and wave with momentum. Light leaves as a particle, travels as a wave and arrives as a particle.

The mathemagical theorist has abandoned the scientific method of inquiry by confusing verbs and nouns, objects and concept, hypothesis with theory, and functional definitions with explanations. They have abandoned science altogether by focusing on behavior rather than architecture.

When their experiments showed them that a particle was impossible, researchers should have abandoned their theories, not the scientific method. When their observations indicated to them that light can not possibility be a wave, they should have abandoned their theories, not their common sense.

Chapter Thirty Two - Light ...Does It Travel Rectilinearly or Curvilinearly?

Let's get something straight between us!

In 'Nature of Light' we cover the light paradox or particle/wave duality as taught to school children and also in institutions of higher learning.

We learn that particle/wave physicists describe rather than explain light behavior because they are confused about its nature. They are confused because they do not have a physical medium for light. We learned that when experimenters didn't see what their hypotheses predicted they invented ad hoc 'explanations' for what they observed. Of course, irrational explanations are not explanations at all.

There is another example of a paradox involving light and gravity. It is the rectilinear/curvilinear duality. Does light travel in a straight path or does it follow the curvature of space? We understand that this is another example of what happens when one doesn't follow the Rational Scientific Method. Why didn't these guys just erase the whiteboard and start over? It would have saved them all a lot of time spent inventing ridiculous ideas like wavicles and geodesics.

One way the particle magician fools himself and others is when he uses ambiguous terms or undefined terms. Does light travel in a straight path? If light is comprised of particles called photons, then no, light would travel rectilinearly along a path. The phi whiz is confusing behavior (travel) with the object (photon). Then of course, this is no ordinary object. The photon has no mass.

Don't blame these poor delusional mathemagicians, they were taught this in fourth grade, and never were able to let it go. Experiments 'prove' after all, and anyways, who are we to question authority?

This nonsense is reinforced later in high school:

Seventeen year old Adnan asks the Department of Physics, University of Illinois:

"Q: Why does light travel in a straight line?

A: A straight line is 'In the eye of the beholder'. As far as light is concerned it travels in a straight line from point A to point B. However, for a distant observer the trajectory may be a bit curved. The reason is that the geometry of space is a bit warped near a massive gravitational source like a black hole or even the sun.

The general phenomenon is called 'Geodesic lines in curved spaces'. [snip] Gravity also acts as a distortion of space; however the mathematics is a bit more complicated."

In other words, light travels in a straight line (rectilinearly) but follows the curvature of space (curvilinearly) caused by objects.

Does that confuse you? Here, maybe this college level explanation from rice college.com will help:

"A line can be described as an ideal zero-width, infinitely long, perfectly straight curve."

Oh that explains it. A line can also be curved. Hence the itinerary duality - straight/curved.

According to wikipedia.com, Eddington "...conducted an expedition to observe the Solar eclipse of 29 May 1919 that provided one of the earliest confirmations of relativity, and he became known for his popular expositions and interpretations of the theory." He was able to photograph a star behind the sun.

Today we know this phenomena as gravitational lensing. This is covered in the chapter on Gravitational Lensing and the CMB.

We are told that objects with mass weigh down the fabric of space and cause light to er..uh..bend..er..uh travel in a curved line around massive objects like the sun. BUT stop the presses! Light doesn't have mass, so it follows the curvature of space. If this were not so, photons would follow curves everywhere there are objects and never travel in a straight line, don't cha know?

Hey! what about Henry Cavendish's torsion balance experiment? It shows that two 348 pound balls attract two 1.61 pound balls. Those certainly aren't massive! Huh...something's fishy here. If small objects cause space to curve, however so slight, it should affect the wee little photon.

There's more! Pound and Rebka were able to detect the effect of gravity on electromagnetic waves in only 22 meters distance from the earth with their Harvard Tower Experiment. Yet, we are supposed to believe that scientists can't detect the effects of earth's gravity on light even though it is traveling a much longer distance to the moon and back!

If General Relativity is right about curved space, then how are scientists able to bounce a laser off of the retroreflector on the moon? This is covered in Rational Science Vol. I in the chapter 'Light, Particle or Wave?' Actually, what is discussed is the particle/wave duality which is clearly debunked by the Laser Ranger Station at McDonald Observatory.

What is also debunked by the Laser Ranger Station is Relativity's curved space. If light is following the curvature of space, how does the laser bounce back to the earth which has already moved a great distance in the time it took for the light to travel there and back! Things are getting stranger by the minute! Maybe we should consider the principle of Ray Reversibility. Wikipedia tells us:

"Retroreflectors are devices that operate by returning light back to the light source along the same light direction."

If space is curved, the light could not have returned "along the same light direction."

We are told that light travels only in one direction unless it is reflected or refracted. The fact that the laser light reaches the retroreflector and returns to its source indicates something that should be obvious: Light travels rectilinearly and in both directions along the same path.

So, as you can see from Pound and Rebka, and Cavendish's experiments, is that we don't need huge distances or massive objects to detect the effect of gravity on light.

Light over distances is covered in the three chapters on "Distance."

The idea of a connecting medium explains light and gravity without the paradoxes. Stay tuned to see what the physical mechanism for light and gravity might be.

Chapter Thirty Three - Distance to the Stars

A friend asked me what I thought about an article on gsjournel.net called...

Distance to the Stars by Jerrold Thacker

Methods of measuring distance to stars are considered and the author states he has a better solution to the current highly inaccurate methods, utilizing the B-V color index. Jerry takes issue with the inaccuracy of parallax technique pointing out that gravitational effects have skewed the results. Instead of stars being billions of light years away, they are more likely thousands, and even closer.

The "measuring" (the magnitude system) has to do with the relative brightness of stars and affixing a distance to that.

Jerry points out, "Without more information it is impossible to tell if a star is very dim and very close, or very bright and very far away. If two stars appear to be equally bright, it is impossible to tell from their observed radiance alone if they are indeed equally brilliant, or at different brightnesses, but at different distances. So a more valuable classification method would eliminate the effect of distance and allow direct comparison."

What does it mean to measure distance, anyways? I'll define using Wolfram scientific definitions:

distance: the property created by the space between two objects or points

measure: the act or process of assigning numbers to phenomena according to a rule; how much there is or how many there are of something that you can quantify

Of course, there's no way to really directly measure the distance to a star, unless we flew there in a spaceship and pulled a piece of string behind us (one that we already had the length of). Maybe we could use balls of string, the distance to our sun: 92,960,000 miles (149,600,000 km) and keep count of how many we use?

Or, is there a space tachometer that can sense space and tick off the miles? Maybe we could drag COBE behind us?

Determining distance by Speckle interferometry and a color index relies on observation. With the limited sensory capability of sight, the distances we measure are confined to the ability to detect light within the visible light spectrum (and color discrimination).

If the speckle technique uses image processing sw base on light as a particle, that would take a different calculation than if based upon a wave. Either way, it seems to me that we would be looking at an average color, a mixing of colors from surrounding target area.

So, the bottom line is we have a relative color, as opposed to, a relative brightness scale.

Another thing...what about the effects of gravity on color? If the author is concerned about the effects of gravity on parallax measurements, why no mention of it relating to color measurements?

Talking about Betelgeuse:

"In fact, it has a measured redshift of 22 miles per second, or a recessional velocity of 79,000 miles per hour. Using a simple calculator indicates that it has receded around 1.3 x 1010 miles since 1993. A very simple explanation for the observed reduction in diameter!"

Redshift is based on the Doppler effect, but what is the medium in space? Sound is not the same as light, it has a mediator - air. What does space have?

Using visible light, Sirius is brightest, using infra red light Beatlejuice is brightest. What does this tell us about distance?

The article's author states:

"Simply put, the B–V index can be used to determine a star's absolute magnitude, and this along with its apparent magnitude,allows the distance to be determined by the formula given below. Although the process is somewhat more complicated than we will give here, the basics are the same. There is a relationship between the B-V index and absolute magnitude for the twenty nearest stars."

To get the B-V color index for a star, we are measuring color intensity. Color is assumed to be based on temperature. Hot is blue, cool is red. The phiz whiz uses blue and green filters to measure the intensity and calculates the difference between them.

Absolute magnitude is the apparent magnitude adjusted to a standard luminosity "distance" (considered as 10 parsecs) from an observer.

Apparent magnitude is the brightness observed from earth and adjusted to remove the atmosphere.The visible light is used as well as (typically) near infrared.

So the B-V color index uses color intensity to determine temperature. The absolute magnitude uses the apparent magnitude, and the apparent magnitude uses an observer (sometimes along with infrared detector device) on earth to determine distance.

So, what we have done is take the difference between color intensity (or actually the 'filtered' difference between the amount, or, intensity of blue and green) and the apparent magnitude to put in our formula, then determine distance. But the absolute magnitude is the difference between the apparent magnitude (observed brightness) and a "standard" luminosity at an arbitrary distance.

In other words, we use a value of intensity than an observer determines, and based on a comparison to an arbitrary luminosity "distance" we come up with one term in our equation (the absolute magnitude). We obtain the difference between this value and the apparent magnitude (which we used to obtain absolute magnitude) minus the atmospheric effect on light. We plug these values into our equation and we get our distance.

And this from Wiki:

"As a star gets cooler and therefore more red, the B-V color index increases, since smaller magnitudes correspond to brighter light."

Cooler temps represent redder color and larger values on the color index, which also correspond to brighter light. Brighter light may correspond to closer proximity in absolute/apparent magnitude, but redder light may mean greater red shift which means greater distance (actually increasing distance in direction of travel away- called recessional velocity).

However both parallax and color index based approaches have inherent difficulties. One has to explain how Doppler Shift in space (redshift) occurs without a medium. Color index relies on absolute magnitude which relies on apparent magnitude. Both are observer based systems relying on luminosity, and an arbitrary 'luminosity distance'.

While gravitational effect on parallax is mentioned, in that it changes the angle of diffraction, Jerry ignores the potential effect of gravity on color.

Would not light bending affect color along with the angle of diffraction? Take a prism, and allowing light to break up into its many colors, project that onto a white surface for easy observation. Now turn the prism ever so slightly (simulating bending light). Notice the changing position (and relative intensity) of the colors on the projected surface. Imagine that you are on that one spot. The color and intensity would change, even though the distance between the prism and the "spot" has not changed.

While Jerrold points out in his articles and books, space is not expanding and there was no Big Bang, he still needs to get over his love affair with Einstein.

Chapter Thirty Four - Shapiro Effect

"You may not have heard of the Shapiro effect before, but you are about to find out that it is the explanation for the distance-redshift effect discovered by Edwin Hubble, and it has been proven experimentally many times!" – Jerrold Thacker; The Deceptive Universe

We are told the Shapiro effect has "proven" General Relativity's time dilation many times "experimentally." Also, that it properly "explains" redshift discovered by Edwin Hubble. Redshift is used to explain the spectroscopic observations of celestial objects recessional velocity, but is an ad hoc position in support of the expanding universe, or, Big Bang Theory.

This is no different than the claim that gravitational time dilation has been proven by our GPS satellites, or, explained by the corrective calculus of relativists. There is a time difference "predicted" by relativity between the orbiting satellite's and the earth station's clocks. What is happening with the atomic clocks is NOT evidence of time dilation, but of a stress caused by varying degree of gravitational attraction on the atoms in the atomic clocks due to a varying distance from the center of the earth to the center of the cesium atoms.

Of course, a similar thing is happening with MIT's Haystack experiments. Experimenters think that the results are confirming relativity's "time dilation" because they can accurately predict the time it should take a radar signal to travel to Mars, "bounce," and return back to earth. This time will vary as the line of site (path) is closer to the sun.

They repeated the experiment showing greater accuracy with Mariner 6 and Mariner 7 Mars space probes, using their transponders. Further test were done using transponders placed on Mars by the Viking project. The fixed position gave even better accuracy. This so-called Shapiro effect, or gravitational time dilation, proves to the relativist that light rays lose "energy" and velocity when passing through a gravitational field, and therefore General Relativity is "correct."

Throughout the previous paragraphs, I have highlighted certain words. Let's take a look at those words as we go over what the astronomer and the phiz whiz is trying to tell us.

If you study the rational scientific method, you will understand that science never proves, science explains. Science never proves: See Proof is for Alcohol.

Experimentation is extra-scientific. Science leaves testing and experimentation to the technicians and engineers. They use a lot of trial and error to build things like GPS, atomic clocks, and radar. The Rational scientific method of inquiry is a means of explaining phenomena with objects, and is purely conceptual. If you believe that proof and experimentation is part of the scientific method then you probably have learned what is being taught as the scientific method. This is anything but science, as explained in my Hub, The Scientific Method ? - For Dummies!", and also Science is Conceptual Technology is Empirical.

When a relativist uses mathemagic to explain phenomena, he is actually NOT explaining at all. He is describing phenomena using abstract concepts, such as energy, gravitational fields, and black holes.

Energy is a magic place holder word, gravity is the attraction between objects (not explained by fields), and the black hole is an irrational abstract mathematical construct invented ad hoc in an attempt to explain celestial objects that the astronomer does not understand. Energy and black holes are covered in detail in my first book of the series, "Rational Science Vol. I" and gravity will be explained later in this book. For now, suffice it to say, gravity is the phenomena of attraction between objects. What the mathematical theorist lack, is a physical medium to explain how this happens.

The particle physicist may attempt to describe these phenomena with particles called gravitons and muons which are purportedly traveling away from their source. In other words, mathemagicians believe that tiny billiard balls smashing into each other are responsible for the attraction between objects!

Time is not an object or thing which can be dilated. Can Jim Croce put time in a bottle? Physics is about physical things. All phenomena are explained through understanding the physical objects involved. A concept is a relation between objects. Time is the relation between locations of objects and requires the memory of an observer. Since science deals with objects, and the Rational Scientific Method removes the observer with his biases and limited sensory system, time is NOT a scientific term.

Rational scientists understand that all phenomena are the result of interaction between objects.

Prediction is a word often used by the mathemagicians as part of their confirmation bias. It replaces the term 'educated guess.' Newton observed an apple fall, and calculating distance and speed, he came up with a measurement, $9.8m/s^2$. Had Newton really been able to predict the apple falling, he would have been able to avoid being hit on the head! The rooster predicts the sun will raise, but if the sun supernovas...so much for the prediction. One simply observes the sun rising hundreds of times, and guesses that it will do so again tomorrow.

As asked in Distance to the Stars:

"Redshift is based on the Doppler Effect, but what is the medium in space? Sound is not the same as light; it has a mediator - air. What does space have?" The answer is nothing. Space is void. Space cannot act as a medium for sound, and it can not act as a medium for light, or gravity! The aether theory was debunked long ago.

Did the Haystack radar signal "bounce" off of Mars and return to whence it came? No, this is a result of the Principle of Ray Reversibility. All experiments with light show that it travels rectilinearly, yet the phiz whiz tells us that this (and the Shapiro Effect) proves Einstein's curved space! Think about it. The earth is traveling around the sun in its orbit at 30 kilometers per second (67,000 miles per hour), and Mars is traveling in its orbit at 24 kilometers per second. If it takes the radar 35 seconds to make its round trip, when it returns the earth would be... (Well you do the math)...a long way away from where it was when the signal left!

This is supposed to prove that relativity is "correct?" Rational science doesn't prove, it explains. An explanation can be rational, or not rational. Correct or incorrect. If it is rational, then the conclusion is possible. That's the best that science can offer.

Relativity is irrational in its proposals of time and space dilation, medium-less gravitational attraction, and energy turn matter, turns energy. Black holes are pure nonsense and were debunked quite thoroughly. Einstein did not claim that his theories predicted black holes. This was Schwarzschild so-called solution (which was a corruption due to David Hilbert) and has been debunked in my book, Rational Science, and also in an article on vixra.com entitled:

"The Schwarzschild solution and its implications for gravitational waves by Stephen J. Crothers"

Conclusion:

Space is not a medium, redshift is impossible. Time is not a thing which can be dilated; black holes are impossible and NOT predicted by relativity. The Principle of Ray Reversibility shows that light travels in a straight path, and since there is no physical medium for the proposed gravitational time dilation to begin with, the Shapiro Effect is an irrational proposal.

Chapter Thirty Five - Distance ...The Rubber Ruler

The Rubber Ruler

Hubble's Redshift stretches space and Shapiro's Effect (gravitational time dilation) stretches time. They both stretch the imagination.

What is space? It is void, nothing, nada, zip; it can't be stretched, expanded, or measured like Play-doh. What is time? Time is motion plus memory; a concept, not a thing which can be dilated, or stretched like a piece of Silly Putty. The mathematician plays by different rules than Mother Nature.

What is distance? It is the separation between objects. If you took a snapshot of the universe, there would be separation between objects. It is static. The mathemagician measures distance traveled, by a single object. This is dynamic.

Relativists define distance in terms of time. The mathematician measures the distance traveled and they measure displacement. Sometimes they are talking about vector quantities, and sometimes scalar quantities. Because they sometimes use a measuring stick and sometimes they use a clock to measure distance it is apparent (except to them) they are not talking about a static distance between objects.

Distance is the space, or separation between two objects. One does not measure space, one measures objects. Time is two or more locations of an object (motion) and an observer's memory.

Relativity puts an observer in the middle of the equation and relies on his sensory system. This is one reason why the mathemagician, theoretical physicist, and cosmologist confuse themselves with things like the Shapriro Effect and Redshift.

Remember the definitions from Wolfram? Here they are again:

distance: the property created by the space between two objects or points

measure: the act or process of assigning numbers to phenomena according to a rule; how much there is or how many there are of something that you can quantify

So as you see, Wolfram attributes distance as a property of space. It can be measured not only between objects, but between points! Those zero dimensional locations on a line.

Mathemagicians can assign numbers to a phenomena based upon man-made rules, and wala! We have billions of light years, or merely hundreds, or thousands of light years between the Milky Way and other galaxies (if they are not illusions, as Thacker alleges), depending on whose rules one chooses.

Where did these poor lost souls go astray? They can't understand the difference between verbs and nouns, objects and concepts, hypotheses and theories, or distance and distance traveled.

The phiz whiz gives space attributes, and makes time a 'thing' rather than a relation. They have chased their tails around in circles with things like length contraction, and time dilation, and space expansion for a hundred years.

Enter Quantum Mechanics (QM), which seeks to unite SR and GR. QM also seeks to unify light, gravity, electricity, and magnetism into one Grand Unified Theory (GUT). Problem is, because they don't understand the most fundamental thing like the difference between an object and a concept, they will never be able to find a single physical mechanism behind these phenomena. There are hundreds of GUTs, and since there are hundreds of mathemagicians relying on their GUT instincts, the phiz whiz has no single physical mechanism for light, magnetism, electricity, or gravity, and they never will as long as they continue along the course they have plotted for themselves.

Rational science uses objects to explain phenomena, where mathemagicians use abstract concepts to describe, or measure phenomena.

Jerrold Thacker and others disagree with BBT's expanding universe, yet fall into other relativist traps. This is because they don't understand the difference between objects and concepts, verbs and nouns, hypothesis and theory, or fantasy and reality. They really need to get over their love affair with Einstein.

Chapter Thirty Six - Relativity's Failed Predictions

From a discussion on Gravitational Lensing and CMB

A gravitational lens refers to a distribution of matter (such as a cluster of galaxies) between a distant source (a background galaxy) and an observer that is capable of bending (lensing) the light from the source as it travels towards the observer. This effect is known as gravitational lensing, and is one of the "predictions" of Albert Einstein's General Theory of Relativity. Please refer to the chapter entitled "Knowledge and Prediction."

Since spacetime around a massive body like a galaxy is curved, light passing by such an object is supposedly bent forming a lens (Einstein Rings). The effect is not restricted to light, and is also being studied in the Cosmic Microwave Background.

Images of Einstein's gravity well always show the sun weighing DOWN space as though it is a canvas. When the sun weighs down the canvas, the earth rotates around it like a ball on a Roulette wheel and then the moon around it.

It's because, as Einstein said, "God doesn't play dice with the Universe." He plays Roulette, and spins it up when you put your money down! Which way is down in space? If the sun is weighing down the space fabric, and earth is spinning around the gravity well, Jupiter would have to weigh the canvas outwards (relative to the earth and the sun) in order for Oberon to travel in the plane that it does facing the sun. AND... If the sun weighs down the spacetime fabric because of gravity, then gravity has nothing to do with the spacetime fabric. Simple logic!

But those aren't the only problems plaguing General Relativity, and the Quantum Mechanics don't have a better solution. Within the framework of quantum field theory, graviton and gluon balls "pull" things together even though they travel away from the source.

A reader, Doc Snow says: "I think you need to look a little deeper into the relationship of math and science.

"It seems to me that your arguments are essentially verbal in nature, while, proverbially, 'mathematics is the language of physics.' Natural language uses metaphor to model the real world--the 'curved space' idea. When you ask "which way is down in space" you take the metaphor 'out of bounds.' The math, on the other hand, remains consistent.

"One last point--you don't actually list 'failed predictions' from the literature. I'm not aware of any, and would be very interested to read about them. (The acknowledged incompatibility of GR and QM isn't a failed prediction, nor are the verbal examples you bring up in the text of your interesting Hub.)"

Sorry, but mathemagic is NOT the language of physics at all. Theorists may wish that were so. Math describes, science explains. Mathematical equations are irrelevant to understanding reality.

Equations can describe phenomena yet explain nothing. Newton's equation described an apple falling, yet could not explain why the apple did not fall up into the sky instead of down onto his head. He admitted that he had no hypothesis:

"I have not as yet been able to discover the reason for these properties of gravity from phenomena, and I do not feign hypotheses. For whatever is not deduced from the phenomena must be called a hypothesis; and hypotheses, whether metaphysical or physical, or based on occult qualities, or mechanical, have no place in experimental philosophy. In this philosophy particular propositions are inferred from the phenomena, and afterwards rendered general by induction." – Newton from Principia

What has come of hundreds of years of mathematics? Not one mathemagician has been able to explain even the most fundamental aspect of reality such as light or magnetism.

As for failed predictions, here are a couple that I mentioned in the article.

SR predicts that clocks run faster due to time dilation (relative to a stationary clock on earth) the further away it is from the earth. GR predicts that the closer to the center of the earth time dilates (slows down). One is relative velocity dilation and the other is gravitational dilation.

SR predicts flat space - GR predicts curved space

BONUS: E=MC^2 FAIL!

Energy is not equivalent to mass. If E=MC^2 and any nonzero amount is added into the formula, as approaching the speed of light, it would take an infinite amount of energy to push an infinite amount of mass.

How can energy (the ability to do work) be equal to mass (a quantity of matter)? How can an ability be equal to a quantity? It can't.

Hence the photon is claimed to be a mass-less particle. Mass-less particles, or point particles, are impossible because zero dimensions are not possible. Otherwise, can you explain how the concept of zero dimensions becomes the three dimensions of reality?

"Let me repeat: "...you don't actually list 'failed predictions' *from the literature.*"

What literature? If you can find anything I have written that is an inaccurate portrayal of "the literature" please feel free to point it out. Of course, if Newton's Principia, NASA, Princeton, Cornell and Physics World, aren't accurate then you are welcome to provide YOUR sources.

"Well, I'm no expert... but you write "SR predicts a flat universe.""

Wikipedia says (and I realize that this is not scientific literature, but it is handy as I write, and it is consistent with my previous reading):

"The theory was originally termed "special" because it applied the principle of relativity only to the special case of inertial reference frames, i.e. frames of reference in uniform relative motion with respect to each other.[7] Einstein developed general relativity to apply the principle in the more general case, that is, to any frame so as to handle general coordinate transformations, and that theory includes the effects of gravity."

Really? That's the literature you wish to use? OK, so, I'll just pick the first sources that verify my point: "special relativity is restricted to flat spacetime" (from WIKI on SR).

What does General Relativity predict about the shape of spacetime near a large mass (eg, a star)?

"What happens in spacetime according to General Relativity, is rather similar. A star will curve and distort the spacetime near it." http://www.hawking.org.uk/into-a-black-hole.html

And yes, Special Relativity isn't about gravity, but it predicts flat space and GR predicts curved space.

But we don't need third party verification, or experiments to understand that the GR theory of gravity is irrational.

I discussed time dilation with the lead scientist who designed the cesium atom clocks on board the GPS satellites. Of course, it was his contention that his calculus proved relativity was correct, because based on Relativity, it accurately adjusted for the difference in time between ground based and sat based clocks due to time dilation.

I explained to him:

It is necessary for a time correction because as the GPS satellite travels an elliptical orbit around the earth there is a variable 'pull' of gravity proportional to the variable distances between the satellite and the center of the earth. This 'stresses the cesium atoms in the atomic clock resulting in variations in time recorded... not time dilation.

A clock's motion through space effects clocks.

Space is nothing. Time is a concept, not a thing. Nothing, and a concept can not be intertwined.

So we have all just seen why conceptually Relativity is irrational. The so-called SR relativistic time correction of 38 microseconds/day is smaller than the accuracy of the GPS clocks. The positional error of the satellite is in centimeters (about 0.8cm!!!) and the GPS accuracy is about 2 meters!

So, even though we don't need it/or use it...even the math doesn't support time dilation for the time variance observed.

From that same article in wiki quoted from above:

"In general relativity, gravity is described using non-Euclidean geometry, so that gravitational effects are represented by curvature of spacetime; special relativity is restricted to flat spacetime."

"Yes, "restricted to" is accurate. If it's *restricted* to flat space-time, then it can't *predict* flat space-time: the former is an antecedent condition--an axiom, if you will."

All math rules are supposed to be axiomatic. SR equations being based on Euclidian geometry HAVE to predict flat space. However, if SR is restricted to flat space and GR is restricted to curved space then one contradicts the other, or neither accurately predicts or represents reality.

BTW, keep in mind SR doesn't have anything to do with gravity, but my reference was about spacetime.

About predictions

The term predicted is used very loosely by the mathematicians. Newton's equation f=ma "predicted" an apple falls so many feet per second square (because an apple fell from a tree and hit him on the head). He measured the distance the apple fell...etc....

A true prediction would have been Newton knowing when the apple was going to fall. According to mathgoodies.com "prediction" is a reasonable guess as to what will happen.

http://www.mathgoodies.com/glossary/term.asp?term=

Newton's math predicts to a high degree of probability that the apple will fall at so many feet per sec. sq. Of course, if a bird catches it on its way down... his prediction fails!

Time dilation is a result of the physical "laws" being used, and is not a law itself. Flat space is an "axiom" because of the use of Euclidean geometry and is being applied to velocity; curved space is an axiom because of the use of non-Euclidean geometry and is being applied to gravity. However, neither are possible because time and space do not exist as objects which can warp, bend, or dilate.

These three are not interchangeable

Newton: one timeframe everywhere

SR: local timeframes

GR: warping of space and time

Relativist love to tell us their experiments prove SR and GR, etc. They use abstract concepts, but they don't have any objects. How can any experiment be done on abstract concepts? It can't! If space and time do not exist, then no experiments can prove that they do!

Continued from wiki article referenced above; so-called predictions:

"Ives–Stilwell experiment – testing relativistic Doppler effect and time dilation"

Relativity predicts gravitational redshift, however without a medium the Doppler effect can't take place.

"Time dilation of moving particles – relativistic effects on a fast-moving particle's half-life"

Time does not exist as in something somewhere; it is a concept that can not dilate.

Hughes–Drever experiment – testing isotropy of space and mass

Space does not exist as in an object with location; it is the concept of separation

"Tests of relativistic energy and momentum – testing the limiting speed of particles"

Energy is a concept. Energy is the ability to do work. Momentum is a concept. It is what something does, not what something is. One can not test invisible objects (particles) by what is happening around them. All phenomena are the result of interaction between objects. We must be able to conceptualize the objects to understand the phenomena.

BONUS:

Why can't Relativists show us a spatially 4-D object? They can't draw one, and they can't even imagine one!

"It's logically impossible to visually imagine a 4D object. That's because we can only ever see any given volume of space one 2D slice at a time." - Physicsforums.com

Although our senses are limited, our ability to conceive is not. Can you conceive of a square circle? NO! Why? Impossible! There are ever only the three orthogonal directions of length, width, and height.

"Experiments to test the aether drag hypothesis – no "aether flow obstruction".

"According to the General Theory of Relativity space without Aether is unthinkable" – Einstein

To find out more about why true/false, right/wrong, proof, belief, evidence, and authority are NOT part of the Rational Scientific Method of inquiry, read the entire series "Rational Science."

Chapter Thirty Seven – What Happened To the Dinosaurs

When I was nine years old I was kicked out of Vacation Bible School for "disrupting the class." The teacher told my mother that I was incessantly asking questions about dinosaurs during the reading of the Adam and Eve story.

After all these years, it has finally been made clear to me what probably happened to the dinosaurs. Today I understand the mechanism behind it, the 'law of extinction.' It reasonably explains why every species eventually goes extinct. It is either background extinction or mass extinction. No massive meteorite collisions with the corresponding impact winter. It wasn't disease, predation, or any of the things that had been postulated and hypothesized before. It was simply the overturning of the population pyramid or inversion of the ecological pyramid.

Amazing that no one had ever proposed the simple explanation of the natural economy until recently. Economics is all about the management of resources, and for nature it boils down to the management of food. Unlike man's artificial economy where three percent of the population provides the food for the rest, other earth creatures have to fend for themselves.

Neanderthal reigned supreme. He was able to gather and hunt enough food to thrive. Eventually, the population grew old and died off by way of background extinction. Background extinction is the overturning of the population pyramid, mostly as a result of density dependent birth rates and loss of genetic stock. The loss of biological diversity is the same for all plants and animals. Declining diversity and increased specialization leads to a collapse of carrying capacity. The young were simply unable to provide for the growing number of old Neanderthals who couldn't carry their own weight. The population inversion resulted in the ultimate demise of Neanderthal. They simply died of old age. It's not the worst way to go, really!

T-Rex, on the other hand, went the way of mass extinction. As older species of plants died off, herbivores depending on those plants starved, and in turn the carnivores that hunted them starved.

During the Cretaceous period, Tyrannosaurus Rex relied mostly on Triceratops who ate mostly cycadeoids and cycads.

Tyrannosaurus Rex, the terrible lizard, was a carnivore. Triceratops, the three horned face, was a herbivore. Cycadeoids and cycads were types of seed plants with strong woody trunks and large stiff evergreen leaves.

Mosasaurs, large marine reptiles, ate mollusks, a type of shell fish such as clams. Mollusks ate plankton which are microscopic organisms like algae and protozoans.

During the Triassic Period, Prestosuchus with his serrated teeth munched on smaller animals like Hyperodapedon, a beaked reptile who scarfed mostly on Dicroidium, a fork-leafed seed fern.

As new plants developed, new species of animals developed along with them. With more plant diversity came greater animal diversity. As older species of plants reached their peak and died off, the herbivores that ate them died off.

The plants conditioned the environment for the next species of plants and animals that inherited the earth. This happened mostly by way of biochemical pumps that created small changes in the atmosphere over a long period of time, usually millions of years. As the atmosphere changed, new plants edged out the old plants, and the herbivores that depended on them starved, as did the carnivores that depended on them.

The nitrogen cycle and carbon dioxide cycles regulate the types of plants that proliferated. Cycads and cycadeoids were crowded out by other newer plants which thrived on the new conditions. Gymnosperms were crowded out by angiosperms. After Triceratops ate most of the cycads and temperatures dropped, new gases and plants arrived on the scene.

Cyanobacteria depended on a symbiotic relationship with cycads. The decline in cycads, led to a decline in cyanobacteria, which led to more decline in cycads. Triceratops populations declined and T-Rex starved.

This same process happened in a number of geologic periods, and this explains mass extinctions consistently.

Chapter Thirty Eight - Expanding Earth Hypothesis
Is it Possible?

Just because the consensus is for Plate Tectonics (PT) does not mean that PT rationally explains geologic evidence (because often the majority of folks are irrational)l. Just because some claims of PT seem unsupported, does not mean that PT is not possible, and just because PT explains our observations does not mean that it is the only rational explanation.

Expanding Earth (EE) makes claims that can be analyzed using the Rational Scientific Method, as does Plate Tectonics (PT). In this article I shall look at the Expanding Earth Hypothesis (EE) to determine one thing: possible or not? Perhaps in a later article I will cover Plate Tectonics, but a hypothesis and its theory must stand on its own.

The hypothesis should introduce the actors in this great play, the language must be clear and concise, and the opening scene should be described in no uncertain terms. Our theory then can consistently and accurately portray the theme, and we can form our own conclusion as to whether we should applaud or not at the end of the closing act.

Jason asks some questions, and raises some interesting issues. Hopefully, we can address these as we go along:

"One thing I don't understand about plate tectonics is why does the oceanic crust spread apart whereas there is subduction at the continental crust? Why not the reverse? I'm also confused as to how the Earth formed with a random supercontinent surrounded by ocean. Stellar metamorphosis would imply that the surface of the Earth should have been uniform, whatever it was. An expanding Earth would explain the continents and the oceans separating them."

As stated previously, the focus on this particular article is, "What does EE say?" We will compare and contrast EE with PT, only in so far as it is useful in clarifying a point. The Rational Scientific Method is only concerned with a particular theory being possible. If both EE and PT are possible, then science is done, and one must decide for themselves which they choose provides a better explanation. There may be third possibility, as well, so of course nothing is ever really settled (unless of course impossible). Therefore, the only thing that need be addressed here is, does "An Expanding Earth explain the continents and the oceans separating them?" Or, more accurately, is it possible for the earth to expand to the degree necessary? We can decide at a later time which theory we wish to "hang our hat on."

"As for matter creation, of course that's not possible. I theorize that the Earth is a gigantic nuclear reactor. Heavier elements are undergoing fission at the core of the Earth, and that process is surrounded by a liquid iron shield. That's what would explain the increase in size without increasing mass."

The real question here is what EE has to say about "the increase in size without increasing mass." In debates and discussions one typically finds "my theory is bigger than your theory" arguments, and avoids all together the question of possibility. Who cares if one theory is better than the other if one (or both) are impossible? Neal Adams claims creation of matter via particle pair production. Impossible!

"I've also pondered whether a smaller Earth would account for the larger prehistoric animals - lower gravity allowing for larger body structures. Of course there could be other explanations, like the composition of the atmosphere, but I still wonder whether gravity would allow for such large creatures. The largest dino is alleged to have weighed over 100 tons, whereas now you have an elephant weighing in at about 8 tons. I'd think a 100 ton creature on today's planet would collapse under its own weight as its legs and feet shattered.

"I asked Bill about this at the conf - would Thread Theory imply lesser or greater gravity on a smaller Earth with the same number of atoms? He replied with Newton's Law, saying a smaller radius would mean more gravity. I then asked if a smaller angle between the ropes would mean a lower gravity, and he said that's possible. I think Newton's Law works when you're talking about the distance between two bodies, however it may not apply to the situation with EE theory. There we have one body expanding, so the radius is greater, but also the angles between interconnecting the ropes increase. The reason radius relates to gravitational strength is the angles, and a smaller Earth means smaller angles, i.e., lower gravity."

The question of dinosaur/gravity size does come up in relation to the EE, so it is a very good question. Whether or not Thread Theory implies lesser or greater gravity has no bearing on EE unless that is a claim being made by its proponents. EE rises or falls on its own merit, so perhaps we'll look at that, in another article and apply Thread Theory to the issue of gravity; "lesser/greater?"

A good response about the dino/gravity issue would be that the dinosaur's muscles were built for today's gravity. We cover this in great detail in another article, "Dinosaurs, How Could Some of Them Be So Big?"

Gravity makes a difference, of course, but so do a number of different things such as, capillary "forces," electrostatic "forces," magnetic "forces," environment, and more. For instance, as discussed in "Size, Does It Matter?" Rational Science Vol. III, insects don't get larger because their respiratory system isn't efficient at a larger scale, neither is their circulatory system, and as they get larger, the rate at which their mass grows relative to their length outpaces their rate at which their strength to length ratio grows. However, this is mostly a constraint imposed because of the exoskeleton rather than the dinosaur's endoskeleton.

Of course, the mainstream idea of gravitation (inverse square of the distance) means Earth's gravitational pull on Dino would be much greater (with a smaller volume). But there were many very large insects 350 to 50 million years ago during Carboniferous and Permian periods. It could be because the planet had more moisture, more oxygen, and was warmer. Oxygen levels are very important for insects, and there may have been 30% oxygen then compared to our current 21% oxygen. But there was also the appearance of birds who love to eat insects. A final point about insects and size is this: There are many insects today that are larger than their ancestors. If a smaller earth is responsible for larger insects, then how does the Expanded Earth explain this?

Here are a couple of other things that the EE fails to address:

Less gravity means less atmosphere. However, water density and viscosity would remain the same, so it could have carried larger particles, and sediments would be increasingly coarser as we go back through older strata.

Apparently there was an earlier (1880's) rendition of the Expanded Earth but the current version was inspired by Sam Carey in the 1930's, and is being promoted today most notably by Neal Adams.

Carey was looking at the continental drift theory of Alfred Wagener, and built a model of a globe (half of one anyways) with continent pieces to visualize what Wagner suggested was happening. The idea of Pangea seemed possible to him (the continental parts fit together) if the earth had been smaller in size in ancient past. This could be accomplished if he shrunk his model, and removed the ocean basins.

He developed his hypothesis with two different types of crust: an older continental crust, and the newer oceanic crust. After the 1950s, Plate Tectonics basically replaced the Expanding Earth as the likely scenario. It appears this is because there is no evidence for an expanding earth, the Plate Tectonic Theory makes "predictions" that are confirmed and the "facts" more accurately align themselves with PT.

Proponents of either side agree that the continents assembled as recently as the Permian and Early Triassic into Pangaea, and that Pangea has since disassembled into the various oceans with new oceanic lithosphere between them. EE has difficulty explaining that there is no oceanic crust older than 200 million years old. PT's answer is that the oceanic crust is pulled into the mantle via the action called subduction. Continents drift, merge and break apart continually (Wilson Cycle).

Although Expanding Earthers often mischaracterize it, according to proponents of Plate Tectonics, the ocean floors are constantly changing, Pangea was not a single continent but an assemblage, and oceans and landmasses assemble, disassemble and reassemble (my words) continually.

Satellites do not show any expansion currently. Expansion is probably an illusion because the earth's surface is shrinking due to subduction. You can compress rock, but it has poor tensile strength. When pushed together it gets thicker and piles up which decreases the surface area. When pulled apart there are cracks which EEers include in their calculations of the earth's surface. They shouldn't do this because it is new crust.

There is no geological evidence that the earth's radius has changed in the last 620 million years, but also, there is no evidence based on what we can glean from using classical physics on the earth/moon system. Nor is there an explanation for how this could be possible. At any rate, periodic variations of thickness of sediments and tidal rhythmites show no such change in moon axial rotation which would be evident with a varied orbit necessary to accompany an earth radius change of such magnitude as required by the Expanded Earth hypothesis.

From the ." - University of Berkely, California:

"Sedimentology offers a methodology for directly tracing the early history of Earth's tidal deceleration, and the evolving lunar orbit through analysis of sedimentary tidal rhythmites… The available Proterozoic rhythmite data are consistent with an overall low rate of tidal friction and the long-term stability of the Moon's orbit.

Most everything we observe is easily explained with Plate Technology, while Expanding Earth has some major flaws. That's what we really want to talk about so …here's one that is insurmountable:

There are a variety of hypotheses. One claims that the mass remained constant and gravity decreased, another that mass increased, and a third that the universal gravitational constant changed. None of these appear to be supported by either evidence or explanation.

Expanding Earth proponents claim that about 100 million years ago the earth was all land, and had to expand to produce the oceans. The earth is one part in three land, which means that 100 million years ago the earth's surface was 1/3 the area.

According to physics stack exchange.com, doing the math for the earth's radius, the earth would have been 1.7 times smaller, and the volume 5.2 times smaller. If density remained the same, as claimed, we'd have a change of mass about 1.5 billion kg/s. Impossible!

The following from NASA factsheet:

Where did this mass come from?

With the mass remaining constant, isobaric compression (pressure) at that volume would make the mantle at the earth's surface about 7300 Fahrenheit, and the crust would melt.

From NASA's Earth Fact Sheet (In case anyone wants to do the math)

Mass	(1024	kg)	5.9726
Volume	(1010	km3)	108.321
Equatorial	radius	(km)	6378.1
Polar	radius	(km)	6356.8
Mean	density	(kg/m3)	5514
Surface	gravity	(m/s2)	9.798

Expanding Earthers can not explain how earth mass can increase, how the gravitational constant changed, or address changing density. Other arguments, such as the African and Indian plates colliding and moving north into Eurasia, pale in comparison.

In the 50's, people thought that if the sea floor is spreading then there are two choices: either crust is destroyed, or the earth is expanding. There was evidence for both at the time. Now we understand more plate tectonics, and about the destruction of earth's crust due to subduction.

CONCLUSION:

The debate always seems to be centered around the driving force of subduction. Proponents of EE must attack this because it annihilates their hypothesis. In my mind, there is no debate until one gets past the major issues that I have already raised about "lack of evidence" or explanation for the earth's radius changing so drastically over about 100 million years ago. Arguments over mantle plumes, ridge pushes, descending slabs, and convection currents are minor issues comparatively, and have little bearing on changes on the size of the earth that is claimed to have taken place 60 to 100 million years ago.

Chapter Thirty Nine - Dinosaurs, How Could Some of Them Be So Big?

Many different theories have been put forth in an attempt to answer this question, and it has remained a paradox to scientists ever since the first dinosaur bones were found. The elephant, at five to seven tons, is a tiny little fellow up against the large sauropods, like brachiosaurus, weighing in at 65-97 tons. Of course, there are no paradoxes in nature. We just haven't been able to understand why some dinosaurs were so big compared to creatures of today.

Related chapter: Expanding earth Hypothesis.

What Happened To the Dinosaur? (Found in Rational Science Vols. I and III and Size, Does It Matter? Rational Science Vols. II and III.

We'll take a brief look at what some of the various hypotheses say, and why it has remained a paradox for so many years. In the chapter Expanding Earth Hypothesis, we covered some reasons why expanding earth makes no sense, and briefly mentioned the Dino/gravity question that comes up often in these discussions. It is claimed that only a lower gravity during the Mesozoic period can account for some dinosaurs being so large.

We'll also address the issues raised in the following article:

Are large dinosaurs proof for Unified field and Expanding Earth? By Edgars Alksnis

"The physical impossibility of the large dimensions of dinosaurs had intrigued thinkers for decades (Desmond, 1976; Bakker, 1986; Lillywhite, 1991; Holden, 1994; Hurrell, 2011). The main objections to existence of largest dinosaurs in today's geophysical conditions can be classified as 1) their bones cannot carry such large weight, 2) their muscles would be not strong enough for movement or flight and 3) their hearts were certainly not possible to pump such large amounts of blood to reach pressures up to 590 mm of mercury (Lillywhite, 1991). The data of biology do not suggest, that Mother Nature might have invented especially strong bones or muscles for this purpose (Schmidt-Nielson, 1984). Similarly, we cannot consider, that the flesh of dinosaurs might be lighter due to reduced water content - data from desert iguanas of present time, for example, show just the same water content than in humans (Minnich, 1970)- some 45ml/100g intracellular and 30ml/100g extracellular water." – gsjournal.net

Some other reasons that scientists think the dinosaurs were able to be so big include:

- The oldest and currently less favored explanation says that reptiles grow throughout their entire lifetimes. "The significant difference between growth in reptiles and that in mammals is that a reptile has the potential of growing throughout its life, whereas a mammal reaches a terminal size and grows no more, even though it may subsequently live many years in ideal conditions" - Encyclopædia Britannica CD (2005)
- Their size was a result of the tremendous amount of available food (vegetation).
- They were big for self defense
- A byproduct of being cold-blooded
- Bird-like lungs
- Egg-laying
- Ate a lot and moved little
- High oxygen levels

Before we cover some of these hypotheses, and then present what I think is the current best candidate, let's get a background on some of the science involved in the "forces of nature," Galileo's Square Cube Law, gravity, density, absolute and relative strength, blood pressure, and more.

Mainstream science tells us that there are four fundamental forces of nature. Rational Science understands that there are really only two, Push and Pull, the chapter, The Forces of Nature, Push and Pull.

Whatever we call them, the "forces" of nature appear to dominate at different sizes. As covered in the chapter mentioned above about size, things do behave differently on different size scales. Not just when dealing with the cosmic scale of galaxies, or the atomic scale of atoms, but whether we are talking about economies of scale, the size of a bridge, or penis size. Size does matter. There is the theory of scale in relativity, and scale is a big topic for art students too.

The so-called strong and weak nuclear forces are related to below atomic level, whereas electromagnetism is said to operate at the level of atoms to stars. Mainstream science also tells us that the gravitational force is too weak to be noticed except where very large objects are concerned. The Earth holds the atmosphere to it and keeps the moon locked into an orbit around it. The sun keeps all the other planets in the solar system in their orbits around it as well.

Whereas gravity dominates the scene when it comes to stars and planets, electrostatic forces dominate in the domain of bacteria.

As discussed in Size Does It Matter, we saw how there is a limit to leaf size for trees, raindrops, and snowflakes have a limit to their size, and gases, liquids, and solids behave differently at different scales.

This is because of something called Galileo's Square Cubed Law. Perhaps, because of the math involved, young children are not taught this in grade school, but we don't need to understand any math to understand the principles involved. Not a single scientific discipline can rationally explain some hypotheses without understanding this important principle.

Galileo wrote about this in his 1638 book entitled, "Dialogues Concerning Two New Sciences." Wikipedia has this to say:

"When an object undergoes a proportional increase in size, its new surface area is proportional to the square of the multiplier, and its new volume is proportional to the cube of the multiplier.

"This principle states that, as a shape grows in size, its volume grows faster than its area. When applied to the real world this principle has many implications which are important in fields ranging from mechanical engineering to biometrics. It helps explain phenomena including why large mammals like elephants have a harder time cooling themselves than small ones like mice, and why building taller and taller skyscrapers is increasingly difficult."

Compare a single one inch square cube with a 10 inch square cube. The area for the 10" cube increased by 100 times. In other words, the one inch square cube is six square inches (1" x 6 sides) and the area of the 10 inch cube is 600 sq. in. However, the volume of the 10 inch cube is 1000 times larger than the one inch cube.

Because of this, scientists became confused with the discovery of very large dinosaurs. They seem to defy Galileo's Law. Persons claiming the Expanding Earth theory often use the dinosaur as a proof of their proposition that gravity was much smaller during the period large dinosaurs roamed the Earth.

Without understanding rules of scale, one may not realize that area and volume of similar objects are not proportional when scaled up or down. Astronomy, physics, and biology can not explain many things without an understanding of this very basic principle.

If the density remains the same with the cubes in our example, the larger cube has a lower area to volume ratio. Since this ratio changes in size, it is critical to understand this, regardless of the scientific discipline one is engaged in.

For instance, paleontologists recognize that there is greater stress on the legs of larger dinosaurs, and engineers recognize greater stress at the base of a larger structure.

For comparative purposes, we find that steel has a tensile strength of 500 and a compressive strength of 500 but bone has a tensile strength of 130 and a compressive strength of 170.

Therefore, while a building may be constructed with concrete, a building twice as tall may require the use of steel which can handle the greater stress. Likewise, larger animals are more likely to break their bones than smaller ones due to the scaling factor also applying to their bones.

Some scientists (Journal of Zoology, Long-bone circumference and weight in mammals, birds and dinosaurs) compiled a chart showing the stress in leg bones of mammals. The Meadow mouse has a standing stress of $1 (N/M^2 E5)$ and the elephant has a stress of 17.1. Did we really need the numbers to understand that there will be more stress on the elephant's legs than the mouse?

Not only do larger animals have lower relative bone strength, they have lower relative muscle strength as we will see in our weight lifting examples next.

Another property of scale relates to absolute strength, and relative strength. Weight is a function of volume, but muscle strength and bone strength are functions of their cross sectional area. A tiny rhinoceros beetle can lift 800 times it's own weight because it has a greater relative strength than a 7 ton elephant who can carry only about a quarter of its weight. This muscle to weight ratio explains why a mouse can jump several times their own height but an elephant can't jump at all. However, the elephant has greater absolute strength, and can lift over a thousand pounds; thousands of times what a tiny field mouse can lift.

An average, healthy man can generally lift his own weight. A healthy 180 lb man can lift about 180 pounds. Of course, weightlifters can do better than that. My friend Fattie's dead lift is 455 lbs. Fattie weighs about 176 lbs. So he is lifting 2.58 times his weight.

The world record guy in that weight class lifts 788.1 pounds and (he weighs 181 lbs.). He lifted over 4.35 times his weight. Of course, this is his specialty, and he can only do this once, but that's another issue.

According to mwolverine.com, the world record for the 275.5 lb class is 899 lbs. Or, the record holder lifted 3.26 times his weight. Yet, the world record for the 114.6 lb class is 564.4 lbs., or, 4.92 times his weight!

This clearly illustrates the principle that relative strength decreases with an increase in size.

Large animals today, like cows, horses, and elephants, have some difficulty standing up from a lying down position. For the large sauropods, like Brachiosaurus, that weighed as much as 97 tons, it would mean that their relative strength would be much less than an elephant one twelfth their size. If one of these brutes fell down, it's hard to imagine how they could get back up.

In Summary: Relative strength decreases with an increase in size; absolute strength increases with an increase in size.

Chapter Forty Two - Dinosaur Size Paradox

In the previous chapter we began talking about properties of scaling related to the ratio of surface area to volume, and covered absolute versus relative strength, and bone and muscle density.

Here's another property attributable to the Square Cubed Law: Since larger objects have a lower surface area to volume ratio than smaller objects, different sized objects have different rates of heat loss. A mouse obviously has a higher surface to volume ratio than an elephant, so they have a higher rate of heat loss than an elephant. For this reason mice have to work at staying warm by eating a lot and huddling together, while elephants have to flap their large ears to stay cool.

For most mammals, a greater amount of the "energy" that they get from their food is used to keep warm than what is used for moving around. The smaller animals eat a lot more frequently than a larger animal to maintain their internal body temperature. A mouse eats roughly 30 percent of his weight every day, while an elephant only eats three to five percent of his weight.

Another aspect of the scaling factor has to do with chemical reactions and diffusion. A large surface area in the lungs and capillary system of mammals speeds up chemical reactions, and allows a greater diffusion rate. In a human body, the blood vessels branch fractally into capillaries so that they cover every square inch of the human body. In fact, there is a capillary no further than a few cells away in any direction. This large surface area allows a high diffusion rate for both blood and oxygen which is needed for the tiny cells (See the fractal foundation.org).

Diffusion rates limit the size of a single cell, affecting metabolic rate, and therefore larger cells are less efficient. Again, this is related to the ratio of surface area to volume. In order to grow beyond the upper limit cells divide.

Take a look at the subject of thermal energy; conduction, convection, and radiation to see how it relates to Galileo's Square Cubed Law. This is important for understanding the smallest unit of life, the cell, or the largest animals like elephants. From the smallest unit of matter, the atom, to massive stars like our sun, size makes a difference on how heat is "transferred."

NOTE: Heat, or temperature, is vibration of atoms and is covered in the book Rational Science Vol. I, Chapter 40, "Temperature What Is It?" Of course, heat is not being transferred at all in that sense, as vibration is what the atoms are doing. An underlying physical mechanism needs to be understood to do the "transferring" or "radiating" of heat.

The terms being used in this article are being used as presented by mainstream science on the subjects covered, and therefore may not accurately reflect the views of the author or conform to principles of Rational Science. It is the intention of the author to present the ideas as they are commonly taught for the purpose of comparing the various competing hypotheses about the size of dinosaurs.

Lots of things affect the transferal of heat, but we are trying to discuss heat as it relates to the surface area to volume (or mass) ratio. A small ice cube melts faster than an iceberg. A mouse cools off faster than an elephant. Temperature changes are faster for smaller objects because the higher surface area to mass ratio.

Animals grow in size from birth to a mature adult, and as they do their requirements change with their size. There is quite a bit of difference between mammals and reptiles in this respect, and this is the subject of "heated" debate; the warm blooded/cold blooded dinosaur debate.

Reptiles, like lizards, do not maintain an elevated internal body temperature like mammals, so a reptile's body looks pretty much the same throughout its lifetime. A human baby, on the other hand, has to maintain an elevated body temperature as they grow, and so their surface area to mass ratio starts off higher than it ends later in life. Not only do babies have difficulty regulating their body temperature, requiring help from their parents, their bodies have to be larger at birth restricting the number of offspring.

I hope that the case has been made for the importance of understanding Galileo's Square Cubed Law because it applies to every scientific discipline. It also comes in handy when looking at the dino/size paradox.

All the formulas for estimating the ratios, bone and muscle strength, conduction, convection and radiation, metabolism and dinosaur mass, etc., are available in standard physics books and on-line so I won't go into those here. Understanding the principles involved is important for conceptualizing the various hypotheses, and this is where we will concentrate our efforts.

The relative bone and muscle strength scaling problem is behind the debate on dinosaur size. Scientists argue that mass of the dinosaur increases at a faster rate than the cross-sectional areas of both the bone and the muscle. Larger animals have less relative muscle strength than smaller animals, as we discovered earlier. They also have less relative bone strength than that of smaller animals. The paleontologist may take the strength of the leg bones and muscle, then divide that by the weight of the dinosaur to determine how much stress it would take to break the bone just standing on it. Then of course, the bones and muscles will need to be much stronger than this to account for walking running, jumping, or falling.

Larger animals are much more likely to break a bone than a smaller animal, but how large is the largest possible size for a land dwelling animal? Whales certainly are the largest animals on earth, but they have the buoyancy of water to support their weight.

There seems to be an upper limit to size today which was not the case during the time of the massive dinosaurs.

Another issue related to the dino/size paradox deals with blood pressure.

Considering raising a column of fluid to a great height, using the relationship between gravity, density and distance, we note that there must be a lot of pressure at the bottom, and that any pump has to be very powerful to compensate for the atmospheric pressure.

Getting blood down to your feet is easy with the help of gravity, but the return trip is a lot more difficult. This is accomplished by the weight of the blood in the arteries pulling blood behind it, one way valves in the veins, and leg muscle contractions.

How can the Brachiosaurus' circulatory system get blood up to his head? A giraffe typically lives about 20 years then typically dies from a heart attack because his heart beats at a very high rate for an animal his size. He has a very large heart weighing 25 pounds, so scientists have suggested the huge dinos would need a heart weighing tons, or multiple hearts, or that they couldn't raise their head above shoulder level.

Why are there so many differing theories, often contradictory ones (like the cold blooded/warm blooded debate) when it comes to this issue? Obviously, there isn't a single known underlying physical mechanism that can answer all the questions satisfactorily. We'll let the scientists continue forever to debate these issues of bone strength, muscle strength, blood circulation, etc., as it relates to dino size. Perhaps you'll research these problems on your own to come up with a solution.

Chapter Forty One - Big Dinosaurs

In the previous two chapters we learned that (because of something called Galileo's Square Cubed Law) as size increases, surface area increases as the square of the multiplier, and volume increases as the cube of the multiplier. For this reason, properties such as heat conduction, density, chemical reaction, and diffusion affect animal muscle and bone density, metabolism, and circulatory systems. This naturally limits their size.

We also learned that an object's (especially a living object's) relative strength decreases with an increase in size, and its absolute strength increases with size.

There are many, often contradictory, hypotheses and theories attempting to explain how some dinosaurs were able to grow so large.

In my thinking, there should be a universal explanation; a physical mechanism that can account for all of these issues at once. It seems reasonable to me that the environment was quite different during the Mesozoic than it is today. From the discussion in "Expanding Earth Hypothesis" you can see it can't be an expanding earth. We'll cover that in a bit more detail, and present a new hypothesis that may just reasonably put an end to the dino-size paradox.

So what changed? It is not likely that it would be bone material. There's no indication that chemical bonding of the elements involved was any different, so strength and density would have remained the same since the Mesozoic Period.

Since muscle doesn't readily fossilize, and there are only a few fragments available to study, we'll have to assume that it too remained essentially the same. Although there is no fossil evidence for it, the Bone Monkeys have postulated hollow bones. This seems unreasonable strictly from the standpoint of evolution.

If hollow bones and stronger muscles were such excellent traits allowing dinosaurs to grow to such large sizes, then it should have been passed on to successive generations.

The Brachiosaurus and modern day elephant may have filled similar niches growing large to protect themselves from carnivorous predators. This can explain why there is such a large gap between the largest and the rest of the animals (and the same for dinosaurs). By comparing the sauropods to the elephants, scientists have arrived at a scaling factor of between 2.3 and 3.5.

Comparing the tallest dinosaur, Brachiosaurus, to the giraffe should give a more precise scaling factor because of the upper limits on the circulatory systems, and the ability to reach higher up for food. Brachiosaurus and giraffe also share long necks, as well as, they have longer front legs than back legs.

Using the scaling factor to determine pressure, stress, bone density, and acceleration due to gravity, one can reasonably estimate that the effective gravity during the Mesozoic Period was around 3 times less than today (3.1 m/s^2).

Without getting heavily into the math involved, we recall the formula $F = G\ M_1\ M_2\ /\ R2$ for determining the effective gravity between two objects. Acceleration due to gravity is 9.8 meters per second squared. The radius of the earth is Equatorial radius (km) 6378.1. The gravitational constant from which G in the formula is derived is 6.67 E-11 N m^2/kg^2 . Msub1 can be the earth's mass ([1024 kg] 5.9726) and Msub2 the mass of Brachiosaurus (97 tons).

What are the possible variables here? Gravitational constant? There is no evidence, nor is it reasonable to assume that it has changed significantly since the Mesozoic. Neither the mass of the earth nor the radius has changed significantly as we saw from the chapter on Expanding Earth.

We'll look at this again from a slightly different perspective. First, let's consider the idea of the gravitational constant changing. A change in gravitational constant would make the universe so chaotic that we would be unable to calculate hardly anything in physics. This would also be effectively shooting William of Occam in the foot. Of course, since there are various differing values for the gravitational constant used, apparently due to noise, different measuring equipment, different experimental parameters, I would say that yes, it does change (wink, wink) but NOT significantly.

Naturally, we wouldn't want to disparage Newton, Cavendish, and many others by changing G, to present a hypothesis on dinosaur size. Most importantly, however, we would hate to be so irrational, because under the Rope Hypothesis, a change in G is impossible!

So let's rule out a change in that variable. What about the mass of the Earth? Early in its formation the Earth grew in mass as it collected space debris; meteors, comets, and dust. However, the Moon (and other planets) can't hide its blemishes and pock marks from us.

Currently, there are only occasional asteroids, etc., that impact the moon or earth. In the case of the Earth, most burn up in the atmosphere never making it to the surface, but it's hard to see what's under the ocean. With no atmosphere, oceans, trees, or buildings, etc., on the Moon we can easily see its crater history going back over three billion years. Let's see… anything major recently? … nope! Let's rule out any significant mass change which would change the effective gravity on the Moon, and therefore vicariously on the Earth (since we share a common orbit and center of mass).

What about the radius of the earth? If the Earth was significantly smaller in the Mesozoic Period, that would make a difference in the acceleration due to gravity. Early in its history, the Earth could have been smaller from compaction related to the moon's tidal forces. This was covered in Expanding Earth so please refer to that article. Also read up on the centrifugal force aspect mentioned below.

The laser ranging stations show that the Moon is moving away from the earth. The Earth/Moon system is moving away from the sun as well. If the earth was half as far away from the moon as it is today, the tidal force was eight times what it is today. Even that wouldn't account for the change in gravitational acceleration necessary for dinosaurs to be able to carry their weight. To produce a gravity of 3.1 m/s^2 the Earth would have to have an average density less than water. Since density must increase as we move towards the center of the earth, this is not rationally possible.

We'll not get into centrifugal and pseudo-centrifugal forces which could affect "Little gee" by pushing up on objects on the surface of the Earth, because although the Earth was spinning slower than today, the buoyancy effect would have been negligible (less than one percent of the gravitational force).

Those are most of the arguments. In the next chapter, we'll cover one individual's startling solution.

Chapter Forty Two - Dino Size Paradox Solved

Atmospheric Buoyancy

Of course, nothing is certain, settled, or solved in science, only rationally explained, and therefore possible. You decide.

In the previous three chapters we looked at various arguments and hypotheses, and potential solutions to the 'Dinosaur Size Paradox.' We rationally explained how significant change in the Earth's radius, mass, and the gravitational constant, is not possible. In this article, we will look at what appears to be a rational explanation for our so-called paradox.

Both air and water are fluids. As such, they provide buoyancy to balloons and to boats. They act similarly, and as one will see later on, some strange things can happen because of it. Perhaps you'll have your own "Eureka!" moment.

According to Britannica: "Archimedes' principle, physical law of buoyancy discovered by the ancient Greek mathematician and inventor Archimedes, stating that any body completely or partially submerged in a fluid (gas or liquid) at rest is acted upon by an upward, or buoyant, force the magnitude of which is equal to the weight of the fluid displaced by the body." - Britanica.com

The buoyancy effect is different in these two fluids (water and air) because different volumes need to be displaced for the buoyancy effect to occur. Water (1000 kg/m^3) is denser than air (1.29 kg/m^3), therefore it takes a greater amount of air than water to achieve the same level of buoyancy.

Land-bound creatures' sizes are limited because of their weight, but sea-bound creatures are not so constrained. Today's whales approach the size of many of the large dinosaurs of the Mesozoic Period.

Therefore, the Earth's atmosphere would have to have a density equivalent (or nearly so) to water in order for the dinosaurs to experience the buoyancy effect.

Yes, let that sink in. This is the hypothesis being presented; Esker's Thick Atmosphere Theory (see dinosaur theory.com).

The Earth's atmosphere had a density two thirds that of water during the period of Brachiosaurus allowing its large body to roam the Earth.

Math-heads can use this formula to calculate if they choose:

"By summing the forces acting on a typical dinosaur such as a Brachiosaurus, the density of the necessary atmosphere is calculated as:

$$\rho_F = \rho_S \left(1 - 1/S.F.\right)$$

where ρ_F is the density of the fluid, ρ_s is the density of the substance submerged in the fluid such as the dinosaur, and S.F is the scaling factor. Inserting into this equation a scaling factor of 3.2 and an overall vertebrate density of 970 kg/m^3, the Earth's atmospheric density during the late Jurassic period can be calculated to be 670 kg/m^3. This says that to produce the necessary buoyancy so that the dinosaurs could grow to their exceptional size, the density of the Earth's air near the Earth's surface would need to be 2/3's of the density of water."

In order to get our minds around this phenomenon of buoyant atmosphere, we need to have a cursory understanding of fluids.

In a nutshell; Earth's atmosphere was formed by out-gassing of water vapor, carbon dioxide, and nitrogen. Nitrogen mostly remained in the atmosphere, carbon dioxide is locked in the minerals, rocks on the ocean floors, and the water vapor became the oceans.

According to David Esker: "It may be hard to imagine that the Earth's air could be so thick that its density would be comparable to water. Nevertheless, there is no reason why a gas can not be compressed so much that it has properties similar to that of a liquid, and in fact compressing a gas into a liquid is a common industrial process.

"In order to compress the air near the Earth's surface, there has to be a substantial amount of overlapping air pressing down on the ground level air. Thus the high density ground level air is evidence of an extremely thick Mesozoic atmosphere.

"Unlike water or other liquids that have nearly constant density between the top and the bottom, the density of a planet's atmosphere increases as one travels from the darkness of space downward to the planet's surface. Close to the planet's surface both the atmosphere's density and the atmospheric pressure is the greatest due to the weight of all of the air above compressing the air at the surface."

To determine the ideal gas pressure at the surface of the Earth during the Mesozoic Period Esker took the ideal gas law equation, and inserted 667 kg/m^3 for density, 294 K for the average temperature, and 43 grams per mol for the atmospheric molecular weight. Based on this, the Earth's atmosphere would have been about 370 atmospheres at the surface.

So in other words, the atmosphere was 370 times what it is today; about what it is at the bottom of the ocean (average depth 3790m - pressure 380).

Dino wasn't crushed anymore than bottom dwelling creatures are crushed today. One needs to understand the difference between absolute pressure, and difference in pressure. As long as pressure on the inside and outside (of a fish or dinosaur) is the same, there is no problem.

Since the thickness of the air was nearly equivalent to water it provided buoyancy to the animals of the Mesozoic, and this allowed them to grow to such large sizes.

Although water absorbs light, and therefore light doesn't so easily penetrate deep into water, light does not have the same difficulty going through the air. The amount of water vapor to gas ratio was small, and so enough of the sun's light made it to the surface.

David Esker's Thick Atmosphere Theory not only explains how dinosaurs were able to grow so large, it also explains the paleoclimate paradox; the planet's climate, and how some tropical plants (and animals) lived at the poles. Instead of the three competing convection current systems (air currents) the Earth currently enjoys, there was a single convection current cell similar to Venus' which more readily moved warm air from the equator to the poles, and thereby kept the climate a more even temperature generating more evenly distributed rain.

I highly recommend David Esker's website, and think his ideas about climate, planetary cosmology, and biology are fascinating and worth investigating. His explanations for dinosaur flight, the purpose of their tails, their mismatched legs, evolution, and formation of the Earth's atmosphere and oceans are worth the read.

I asked Esker about this from Dikran Marsupial at Skeptics stack exchange.com: "volcanic pumice has a density of about 640 kg/m3, so if Jurassic air density was 670 kg/m3, then pumice would float... in the air! As would a lot of vegetable matter, such as any wood less dense than mahogany, which rather suggests that Jurassic air density was perhaps unlikely to have been that high!

"Hi [Monk],

"In the present world it would seem very odd to see low density rocks such as pumice floating in the sky. Likewise it would be even stranger to witness giant reptiles as large as a giraffe such as the Quetzalcoatlus taking off from the ground and flying about. Likewise it would be equally strange to see animals the size of whales such as the Brachiosaurus walking about. So Dikran Marsupial's point is …that he has no imagination? Yes, things were different during the Mesozoic era. That is what the evidence indicates.

"While most of the time, the thick atmosphere would be clear of floating volcanic pumice occasionally this interesting phenomena would likely occur during Mesozoic era and at other times when the Earth had an extremely thick atmosphere. No big surprise there since the same sort of thing will sometime occur in today's oceans when volcanic activity takes place below the waves.

"During times when the atmosphere was thick, an extremely large release of light volcanic material would have the potential of blocking out the sun for an extended time and this could produce a mass extinction. On a few occasion this probably happened. Scientists are slowly coming around to recognizing that volcanic activity is probably the cause of most if not all of the major mass extinctions.

"The thick atmosphere theory is a mind blowing revolutionary idea in the same way as Copernican heliocentric theory was a mind blowing idea centuries ago. In both cases, the evidence supporting the new idea was/is overwhelming, while the previous belief was/is extremely stupid. Hence, if all scientists were as unimaginative as Dikran Marsupial there is a good chance that we would still believe that the world is flat." - David Esker

Chapter Forty Three - Mach's Principle

Mach's principle According to Newton, Mach, and Einstein

Einstein coined the term "Mach's Principle" relating it to gravitational theories, but there are many aspects of Mach's principles. Accordingly, we are told, even Mach himself was never clear what his principle was. In fact, Einstein referred to it as an "imprecise hypothesis."

The idea, principle, or hypothesis may be generally considered as this: locally, inertia is determined by the distribution of matter cosmically. In other words, physical laws that govern the motion of stars and galaxies Far, Far Away also insure that you will experience the centrifugal force here.

While it makes sense to relate the motion of one body to the motion of other bodies, it is more than a philosophical matter. Thought experiments with a single rotating body make the idea of motion meaningless when one understands that space is NOT an object.

Einstein likely was intrigued by the idea of inertial frames while developing his general theory of relativity. However, unlike "time" which has no physical presence "in space" or any relationship to objects, motion is only relative.

Objects may drag other objects around, but objects can not drag the concept spacetime around.

Whereas Newton's gravity depends solely on the mass of an object, the Lense-Thirring effect, within the ToR predicts that a large rotating body would distort spacetime affecting nearby objects. Newton's law of gravity does not predict frame dragging, as the effect is too small to be seen with small objects. One can only observe frame dragging as it relates to very large objects on a cosmic scale, like stars and planets. While size does matter, and phenomena does vary based on size of the objects involved, there is a single underlying physical mechanism which governs all phenomena (see Rope Hypothesis). NO need to invent nonsensical concepts interacting with objects and other concepts.

Other applications such as the Gödel rotating universe do not follow Mach's Principle. We do not, however, need to consider such nonsense as this or Einstein's metric tensor, as these are mathemagical abstractions having no relationship to reality.

We need only discuss briefly the question of absolute or relative motion. It would be more than prudent of us to define the Key Term motion: Two or more locations of an object

What could absolute motion ("without relation") even mean?

Mach said this:

"[The] investigator must feel the need of... knowledge of the immediate connections, say, of the masses of the universe. There will hover before him as an ideal insight into the principles of the whole matter, from which accelerated and inertial motions will result in the same way."

Indeed, if all masses (units of matter) are interconnected, inertia and acceleration "will result in the same way", that is by the same mechanism.

Mach criticized "Newton's bucket argument" in his "The Science of Mechanics." He took exception to the idea of absolute space. Newton laid out the argument in his work "Philosophiae Naturalis Principia Mathematica." It goes something like this:

A person always knows when they are rotating with respect to absolute space by measuring the forces. Fill a bucket with water and observe as the bucket is rotated. The water starts off still and then begins to move up the sides of the bucket because of centrifugal forces. Absolute space in this instance is earth's "reference frame", but distant stars can also be considered. When the bucket is rotated with respect to the water we see no such centrifugal effect since the bucket is now also moving with respect to absolute space. Of course, water and bucket are moving with respect to the observer and also to the earth and distant stars.

Keep in mind that Newton had the good sense to admit this:

"It is inconceivable that inanimate Matter should, without the Mediation of something else, which is not material, operate upon, and affect other matter without mutual Contact...That Gravity should be innate, inherent and essential to Matter, so that one body may act upon another at a distance thro' a Vacuum, without the Mediation of any thing else, by and through which their Action and Force may be conveyed from one to another, is to me so great an Absurdity that I believe no Man who has in philosophical Matters a competent Faculty of thinking can ever fall into it."

According to the Wiki article the bucket experiment could only inform us with respect to centrifugal forces if the bucket were "leagues big." Therefore Mach called for the concept absolute motion to be replaced with relativism where motion of a body is considered with respect to all other bodies.

I'm not sure what Newton meant by absolute space in light of his statement about an underlying physical mechanism mediating action at a distance. Perhaps he meant "all objects are interconnected, but I do not understand by what, so I'll call this absolute space." Perhaps in trying to work this out he considered ether as did others of his time and before him:

"Doth not this aethereal medium in passing out of water, glass, crystal, and other compact and dense bodies in empty spaces, grow denser and denser by degrees, and by that means refract the rays of light not in a point, but by bending them gradually in curve lines? ...Is not this medium much rarer within the dense bodies of the Sun, stars, planets and comets, than in the empty celestial space between them? And in passing from them to great distances, doth it not grow denser and denser perpetually, and thereby cause the gravity of those great bodies towards one another, and of their parts towards the bodies; every body endeavoring to go from the denser parts of the medium towards the rarer?" - Isaac Newton, The Third Book of Opticks

Einstein also believed in the aether: "We may say that according to the general theory of relativity space is endowed with physical qualities; in this sense, therefore, there exists aether. According to the general theory of relativity space without aether is unthinkable;

for in such space there not only would be no propagation of light, but also no possibility of existence for standards of space and time (measuring-rods and clocks), nor therefore any space-time intervals in the physical sense. But this aether may not be thought of as endowed with the quality characteristic of ponderable media, as consisting of parts which may be tracked through time. The idea of motion may not be applied to it"

Newton rationally suggested an interconnecting media, but had no hypothesis for it. Mach also considered that all masses must be related to one another. He may have envisioned an underlying physical mechanism. The rope model explains Mach's Principle, as does this equation: f=cλ which is the equation of a rope. Einstein's Special relativity denied aether, and yet his General Theory of Relativity required that space be a medium.

What Newton, or Mach, or Einstein really thought is not clear and is debated to this day. As the wiki article points out:

"Because intuitive notions of distance and time no longer apply, what exactly is meant by "Mach's principle" in general relativity is even less clear than in Newtonian physics and at least 21 formulations of Mach's principle are possible, some being considered more strongly Machian than others.[8] A relatively weak formulation is the assertion that the motion of matter in one place should affect which frames are inertial in another."

Attempts to make sense of Mach have led to many a theory "more Machian" but none have been able to cut the theoretical mustard.

In other words, if we throw out all rationality we can't make hide or hair of Newton or Mach's centrifugal force, but it may have to do with motion of bodies affecting motion of other bodies. This is an inherent problem of all of theoretical physics: "We can't tell the difference between objects and concepts, but if we finagle the math well enough we can predict anything."

Frame dragging, spacetime, fixed background, asymptotically flat universe, inertial frames, elliptic partial differential equations, energy-momentum, spatially compact and globally hyperbolic, are all descriptive of they know not what.

Here is a treatment of the absolute/relative space/motion arguments, in terms of its history from an article entitled Mach's Principle, by Herbert Lichtenegger, arxiv.org:

"The absolute view identifies space with a container holding all material objects in which the bodies can move but which exists independently of its content, while the relative view considers space merely as a conceptual abstraction of the storage of the bodies and is thus based on the existence of bodies, losing its meaning without them."

Interesting analysis, but I take away something quite different from the typical arguments from theoretical physics about space. Space is always some"thing" that can bend, warp, and ripple. AND as is typical, I could not find a single definition for motion, or for space!

Mach's Principle, and f=cλ, as in all "principles" and equations, are trustworthy in so much as they can be explained by the underlying physical mechanism. IOW, we don't try to force Rope Hypothesis to conform to these formulas or equations.

Many similarities are seen, or connections made, such as that between electrostatics and gravity (both have attractive force), or Mach's Principal being an inherent part of Maxwell's equation (both being vector field theories). Mach's principal, or conjecture, is often referred to as imprecise, or too general.

Both gravity and electrostatic "force" are stronger or weaker based on "the inverse of the distance squared." Even the two have similar formulas:

$$Fg = G.M1.M2/D^2$$

$$Fe = K.Q1.Q2/D^2$$

Any similarities or connections are a result of the inter-connectedness of all matter. The issue is the lack of a single physical mechanism for gravity, electricity, magnetism and light. Well… not an issue for Rope Hypothesis.

More on gravity and electrostatics to come. Don't miss an advanced discussion of the topic in my book Rope Hypothesis and Thread Theory, coming soon to Amazon!

Chapter Forty Four - Atoms
What Do They Look Like?

No one knows. We can only conceive of their structure because it is impossible to see an atom. A hydrogen atom is supposed to be 0.1 nanometers (nm) yet the visible light spectrum is 3800-7500 angstroms, or, approximately. 400-700 nm.

It is impossible to see an atom since all atoms are thousands of times smaller than the smallest visible light 'waves' that we can see. In other words, atoms can't be imaged by any optical system.

What is an atom?

According to Wikipedia:

"The atom is a basic unit of matter that consists of a dense central nucleus surrounded by a cloud of negatively charged electrons. The atomic nucleus contains a mix of positively charged protons and electrically neutral neutrons (except in the case of hydrogen-1, which is the only stable nuclide with no neutrons). The electrons of an atom are bound to the nucleus by the electromagnetic force.

Atoms can only be observed individually using special instruments such as the Scanning Tunneling Microscope."

OK, more on those imaging devices later. First let's define the term atom.

Webster's Dictionary:

1: one of the minute indivisible particles of which according to ancient materialism the universe is composed

2: a tiny particle : bit

3: the smallest particle of an element that can exist either alone or in combination

Wolfram Scientific Dictionary:

Atom; physics and chemistry: the smallest component of an element having the chemical properties of the element

Alright, the ancients proposed that the material universe was composed of particles they called "atoms" as the fundamental, or, smallest part. Of course, science should only be concerned with the material universe - ancient or modern.

We are told that the atom is a particle and that it is the smallest component of an element. However, the atoms themselves are said to be comprised of still smaller parts such as protons and electrons. The structure of the atom and the inter-relating parts vary depending on which atomic model is being proposed.

Let's take a look at some of the more popular models that have been proposed.

1. Thomson (Plum Pudding) model - replaced his nebular atom

2. Rutherford Planetary Bead - replaced Thomson's Plum Pudding

3. Bohr Planetary Bead - an improvement, or, quantum interpretation of Rutherford's model

4. Sommerfield's Wavon - incorporates Relativity as an improvement on the Bhor model

5. Debroglie's Ribbon, or, wave model of the atom; an improvement on Sommerfield's model describing electrons as acting as both a particle and a wave

6. Schrödinger Wave Model - a probability function or cloud region where electrons are likely to be found

7. Born's Electron Cloud - used to find the probability of a single or a pair of electrons in any specific region

8. Lewis Shell - diagrams used to show the bonding between atoms and molecules or pairs of electrons that may exist in the molecule

Each atomic model is used to address issues with previous models. Physicists describe models that make certain 'predictions' and then perform experiments which contradict their models. Instead of scrapping failed hypotheses or theories and starting over, they amend them.

There are no paradoxes in nature such as particle/wave duality and this is the root of their problems. Please see the chapters in this book and also in Rational Science Vol. I on light. Theoretical physicists are using math to describe how the atoms behave. They are not actually describing how atoms look. Theoretical physicists fail in the hypothesis stage when they do not define their terms unambiguously.

Many of the phiz whizzes concepts such as electrons, protons, neutrons, and the various Standard Model particles that atoms are comprised of, are zero dimensional! Please see my articles on dimensions to understand why this is not possible. The Rational Scientific Method requires that you illustrate all the objects of your hypothesis. These particles are hypothesized in many models by the establishment. Not a single model has a photograph or image of their proposed atom.

So what are they looking at?

There are various methods of imaging the "atom." The two most common devices currently used are the Atomic Force Microscope (AFM) and the Scanning Tunneling Microscope (STM).

Atomic Force Microscope scans the surface of an object with a needle tip. This device is basically "feeling" the surface or in non-contact models, detecting electric potentials.

Scanning Tunneling Microscope (STM) uses QM tunneling currents to detect electron densities. The STM resolution is around 0.1 nm lateral resolution and 0.01 nm depth resolution. So, we are looking at "surfaces."

How does a cloud, or, region provide a surface? How does one locate the surface of a probability? Where is the surface, on a wave, or on a particle? Clearly, we are not actually seeing the surface of an atom, but composite images of something, we know not what.

Atoms are said to be made of protons and electrons. Protons and electrons are said to be comprised of smaller particles including Quarks. Quarks are one of the basic building blocks of everything in the universe including protons, neutrons, and other subatomic particles that make up the nuclei of atoms.

The idiots at Fermilab would have us believe that the top quark weighs more than the gold atom. Over 400 scientists received the Nobel Prize for this 'discovery.' Right! The part weighs more than the whole!

Its obvious to me, we need to erase the whiteboard and start over. We need an alternative to particle physics, which explains light, gravity, electricity, and magnetism without the contradictions or paradoxes of particle physics, or the nonsense of Quantum Magic.

To quote a friend: "Not a single atom has been seen. In fact, it is impossible to see an atom because the atom is what initiates light signals. You cannot use light to see an atom."

Chapter Forty Five - Planetary Evolution

Part One

Cosmology is the study of the "origin and evolution of the universe." "Cosmogony is any model concerning the coming-into-existence of the universe or sentient beings" and a primary question is, "How do planets and stars form?" There are a couple opposing ideas about this, but only one is typically taught in schools and institutions of higher learning:

The nebular hypothesis.

In this series of chapters we'll take a cursory look to see what we can find out.

Of course, as we review material, we'll have to pick the sticks out of the hay before we eat it. There are certain core principles, like the mechanisms of gravity, electricity, magnetism and light that form a grid these theories will have to fit through. Let's try to be open minded without letting our brains spill out onto the keyboard.

I'll assume that the reader is familiar with the ideas of the Rational Scientific Method, and avoid lengthy discussion about foundational principles of physics. There are very many issues which have been repeatedly addressed in my previous articles, books, and at the Rational Scientific Method Facebook group on space, matter, and time, etc. which can not be covered here due to the scope of the subject and brevity of the chapters.

For example: In many discussions an important question, "What is the universe?" is usually avoided. When one looks up the definition of universe, this is what they are likely to find: The Universe is all of time and space and its contents. This raises questions that are never really addressed, such as, "If space is needed as a container, what contains space?"

The answer to "Did it have a beginning and will it have an end?" is often assumed in the definition of cosmology: WIKI says that cosmology is the study of the "origin, evolution, and eventual fate of the universe." However, there is no reason to assume that there was a beginning to matter and motion, and every reason not to accept this premise. Yet Big Bang Theory depends on creation of matter, space and time. Whatever happened to steady state? We'll take a cursory look later.

I'll make it easy on my self and everyone else by just looking at several opposing views: nebular hypothesis, and stellar metamorphosis, or the more coherent transformative metamorphosis.

The mainstream teaching of nebular hypothesis is consistent with the Big Bang Theory (not the TV sitcom, but just as hilarious) and hence "origin...and eventual fate" in the definition of cosmology. The reader is advised to read through the previous articles presented in the book series Rational Science Books I through IV, or visit the Facebook group's files section and thread archives here.

This way one can glean a basic understanding of the arguments for and against the Big Bang, or Steady State, as well as learn something about the nature of space, time, matter, gravity, light, electricity and magnetism.

Nebular hypothesis is taught exclusively in elementary schools, colleges and universities world round, so I will only briefly outline it here and then we can compare and contrast it with other hypotheses as they come up.

Nebular hypothesis

Kant developed it in 1755 in his paper, "Natural History and Theory of the Heavens" to explain how the solar system was formed and evolved. Now-a-days cosmologists believe the Nebular hypothesis is how all planetary system develop throughout the Universe, although, most of what Kant wrote has been replaced.

The theory attempts to explain circular orbits and why planets move in the same direction that the sun orbits. Stars are formed from large, dense, gaseous, molecular hydrogen-helium dust clouds which merge forming clods, which begin to rotate, condense and form stars.

A protoplanetary (accretion) disk which forms around the star feeds the star as it siphons off its mass over about a million years or so. After the disk cools off it forms dust grains which can become planetesimals. If the disk is large enough in a few thousand years we may have a planetary embryo. Over the next 100 million to a billion years these embryos merge into a few terrestrial planets.

How giant planets form is a bit more complicated and may occur behind the frost line. This is where little bitty baby embryo planets are conceived mostly of ice.

Some babies grow up to be earth sized planets reaching a threshold needed to suckle hydrogen and helium gas from the accretion disk. This takes millions of years, but when the young proto-planet reaches a size 30 times that of earth the hungry planet begins to scarf that gas at a much faster rate where it gains the bulk of its mass in about 10,000 years.

When the proto-titty goes dry the full grown planet now takes off on its own and roams often far from home. Some, like our own Uranus, become icy planets when their core fails because they formed at a time when the gas disk was almost exhausted.

Opponents of this hypothesis point out what they consider fatal flaws:

Gas typically expands, not coalesces. In order for a star to form where enough pressure initiates fusion in its core the gas cloud would have to be very, very large. The gas would have to be very dense in order for gravity to overcome gas's natural tendency to expand. At some point, called the critical mass, gravitational collapse would overcome the expansion rate of the gas. Critical mass is a function of gas density and temperature and has a formula created by mathematician James Jean.

A dense cloud favors collapse and a hot cloud favors expansion. The Big Bang Theory (BBT) says that the temperature at the time of star formation was very high, and it would require a cloud the size of a hundred thousand suns. No stars could have formed from any gas cloud less massive than a globular cluster.

The sun has an axial tilt greater than it should have, since it should be spinning in the same plane as the other planets in the solar system. Its axis is tilted over 7 degrees from the ecliptic. Collisions may explain the tilt of planets, but not the sun's.

Why are there retrograde rotations like Venus? As the nebula formed and spiraled inward, all the planets should orbit and spin in the same direction.

Our sun's spin is too slow. Angular momentum is equal to mass times velocity times the distance from the center of mass, and must remain constant in an isolated system. The smaller our sun got as it collapsed, the faster it needed to spin in order to conserve angular momentum. Since the sun has 99 percent of the mass of the solar system and only 2 percent of the angular momentum, it appears opposite of what nebular hypothesis predicts should be the case.

In the next chapter we'll look an alternative hypothesis: Stellar Metamorphosis.

Chapter Forty Six - Stellar Metamorphosis

Planetary Evolution Part Two

As an alternative to the nebular hypothesis, we'll take a look at Stellar Metamorphosis by Jeffrey Wolynski. His articles can be found on vixra.org.

Any hypothesis or theory should stand on its own merit, and not be discarded by virtue of its opposition to mainstream hypotheses and theories. Indeed, they all must be held to the high standards of scientific inquiry as proposed by the Rational Scientific Method.

I downloaded a couple of Jeffrey's introductory papers. I couldn't find terms defined anywhere in one and only these unscientific definitions on page 7 of the paper linked to above Entitled, "Stellar Metamorphosis: An Alternative for the Star Sciences." Sub-Subtitled, "A planet is a star and a star is a planet."

• Solar system, star system (n): an area in space where at least two stars orbit each other

• Eclipsing binary stars: (n.) Young solar system/young planets

• Exo-planet (n.): A star/planet that does not orbit the sun

• Star (n.): New planet

• Planet (n.): Old star

• Planetary system, star system (n.): any area of outer space where at least two stars/planets orbit each other

• Proto-planetary disk, debris disk, Circumstellar disk (n.): field of shrapnel from star collisions

Let me remind the reader that a scientific definition must NOT be: circular, synonymous, ambiguous or contradictory.

Wolynski calls for creation of matter, hails the Electric Universe and plasma, and makes unsupported claims like Pulsars pulse because they switch back and forth between being inductors and capacitors. If Wolynski can't explain the underlying mechanisms of electricity, magnetism, and gravity, how can he claim that electricity and magnetism, rather than gravity, are key in star formation? Well, he doesn't and he can't, at least with any sort of scientific integrity.

Jeffrey was all too quick to point out inconsistent definitions of mainstream, yet provided some unscientific definitions and used electromagnetism, plasma recombinations, Murkland convection, Birkeland currents, and current density, without defining them at all, and in fact suggests to the reader to study up on such things as the Meissner Effect "Because star formation may require it."

Indeed, his "hypothesis" does require the use of such terms as Z-pinches, electric and magnetic fields, radiation, and other such mainstream terminology.

The article says things such as "Gravity is not a constant" and "electromagnetism births stars" without ever explaining either electromagnetism or gravity. Wolynski does say that electromagnetism is responsible for gravity, and as our sun cools, "other cooling stars (planets in the SS) will loose their orbits because the sun's electric and magnetic fields will continue to lose their reach." This may have something to do with angular momentum as JW suggests in the same paragraph that we read up on it.

How are we to take Jeffrey Wolynski's love/hate relationship with mainstream science when he picks and chooses from them to suit his so-called hypothesis? And how are we to respect his charges against mainstream and the Electric Universe for inconsistent definitions when he doesn't provide scientific definitions of his own?

It doesn't bother me when I hear such things as Gravity ... can not "accrete material whatsoever" and "Gravity is not a constant." It just gets my knickers in a knot that JW doesn't explain what gravity is.

Without an illustration of the physical mechanisms of gravity and electromagnetism, when I hear, "current density is probably the effect that is known as gravitation" I have no clue what he is talking about.

And when I hear things like, "A star that dies must get rid of its charge separation." I think of mainstream ghosts and massless particles, and other such reified and refried nonsense.

One of his followers who has his own video about stellar metamorphosis acknowledges, "It essentially is Oparin's. It is a further development of Oparin's work with additions." Oparin speaks of planetary evolution as though it had been previously considered. Rightfully so, as we shall next in our series of articles.

When discussing the hypothesis with various proponents, including Wolynski, when I asked for definitions for electricity, magnetism and gravity, or ionization and energy, etc., the typical reply is as follows:

"The terms you listed I see as mysteries. We still don't know everything there is to know about them. I'm still scratching my head on electricity."

No definitions of Key Terms, therefore no hypothesis, therefore no theory yet. Keep working on it Jeffrey, and get back with me in another couple of years.

Chapter Forty Seven - Transformation Hypothesis

Part Three Planetary Evolution

Anthony Abruzzo's Transformation Hypothesis can be found at gsjournal.net.

His 7 papers give us a history of planetary evolution hypotheses and they point out the main objections to nebular, or accretion hypothesis and its variations. Abruzzo offers up an alternative to what he calls derivative hypotheses. His view is one of transformation not derivation, or in other words, one star becomes one planet, instead of stars and planets forming together as a solar system.

Descartes published Principles of Philosophy in 1644 presenting a vortex cosmology that influenced such scientific minds as Huygens, Hooke, and Liebniz. This thinking prevailed for over a hundred years until Newton combined earthly and heavenly physics.

Descartes' vortex cosmology dealt with the stellar formation and migration of planets. In other words, he envisioned stars being transformed into planets.

Newtonian gravitational attraction took precedence in the explaining of solar system formation as persons such as Buffon and Kant championed the catastrophic and derivation hypotheses. A nebulus cloud with a central body forming due to gravitational attraction, condensing dust and gas; two celestial bodies such as two stars; or a star and comet resulting in smaller bodies orbiting a large central body, became the assumptions of various derivational hypotheses.

So we have the elegant Cartesian evolution of one star becoming one planet, loosing favor to various Newtonian derivative hypotheses.

Abruzzo takes the time to compare and contrast today's conventional stellar evolution to his Transformation Hypothesis, pointing out the obvious problems. For instance, since cosmogonists rely on Big Bang Theory's 13.7 billion year universal time frame, we can only predict that our sun will become a black dwarf. There simply has not been enough time for any star to become a black dwarf. The star must first pass through the various stages of fusion, red giant and white dwarf. It must expand and contract repeatedly as gravity and fusion battle it out while loosing half of its mass.

A black dwarf has dissipated all of its heat, fused its hydrogen and helium, and lost its mass to the solar wind, CMEs and flares. A red dwarf has depleted all of its core hydrogen, but some remains in outer shells and can still produce helium. As a result of the fusion moving outward the star gets much larger, the helium in the core becomes carbon and oxygen and more energy is released. When the hydrogen in the shell runs out the sun contracts until pressure from helium fusing causes it to expand again resulting in a chain reaction in the next hydrogen shell, and the star expands again. Expansion and contraction continues until helium and hydrogen are expended.

The mass being ejected from the red dwarf forms a planetary nebula around the star.

Nuclear fusion stops and the star has entered into its white dwarf stage where gravitational forces causes the star to collapse to a dense object about the size of the earth but much denser.

It has a much higher surface temperature than in previous stages, but lacking fusion at its core will dissipate its heat. Eventually it stops loosing mass and slowly cools into a small, dense, invisible black dwarf potentially with some planets orbiting around it.

One particular model calls for the sun never quite cooling completely because of something called proton decay. It will continue radiating thermal energy and losing mass until it completely disappears leaving nothing behind but some gamma rays. Eventually, the entire universe will disappear if one take of the Big Bang Theory is correct.

Proton decay could explain how stars become planets in the transformation hypothesis.

Cosmologists use a spectral classification dividing stars and planets into 7 different classes (O, B, A, F, G, K, M) based on their mass: brown dwarf, gas giants, rocky planets, dwarf planets like the moon and objects in the Kuiper Belt. There is some overlapping between categories and the International Astronomical Union has adopted these terms: brown dwarfs, planets and dwarf planets. There are thirteen orders of magnitude between the smallest and largest category. Abruzzo imagines that it is no stretch to assume the largest being the youngest and the smallest the oldest evolutionary object. Black dwarfs being the oldest.

Of course, even with proton decay being assumed, the universe would have to be much older than 13.7 billion years for there to be any black dwarfs.

Anthony Abruzzo avoids getting into a debate between Steady State's eternal universe and Big Bang's 13.7 billion years and instead takes the "middle ground" opting for a universe of an indeterminate age. This allows him the luxury of avoiding any time restriction of BBT and also to accommodate the proton decay hypothesis to account for mass loss.

Since neither Steady State's eternal universe nor Big Bang's creation universe can be proven, says Abruzzo, it can not be scientifically analyzed. Other arguments for and against an determinate age are made with the idea that transformation hypothesis "requires more time than allowed by the Big Bang Hypothesis but less time than the eternity contemplated by the Steady State hypothesis." Anthony Abruzzo says that either position is "colored by metaphysics" and to question whether the universe had a beginning or not falls "outside the purview of science."

The first article is summed up like this:

- Planets should be viewed as "end products not by-products of stellar evolution."

- Overlapping masses between classes implies "an evolutionary continuum."

- Loss of mass and contraction occurs simultaneously.

- Proton decay could account for the gamma radiation we observe, and also can account for the cosmic microwave background.

- In 1994 terrestrial gamma flashes were detected in the earth's upper atmosphere, and it is supposed that the mechanism for this is proton decay.

- There may be other mechanisms for mass depletion not known at the present.

- Whether or not the universe had a beginning should be left to the metaphysicians.

Chapter Forty Eight - Solar System Formation

Part Four, Planetary Evolution Transformation Hypothesis.

Transformative hypothesis sees planets as end products versus by-products of nebular hypothesis. So rather than planets forming from proto-stars, planets are a later development in the evolution of stars.

Continuing on with the second in a series of papers by Anthony Abruzzo's, we'll discuss in greater detail some of the problems of what he calls derivation hypothesis and see what transformation hypothesis has to say about it.

We mentioned in Part One that a problem for Nebular hypothesis is angular momentum. Why is the sun spinning slower than predicted? Why does it have 99 percent of solar system mass, but only 2 percent of the angular momentum.

Hannes Alven, and later Fred Hoyle, proposed that during the initial stage of the sun's formation something called magnetic braking transpired and this has been adopted by many in the current crop of planetogonists. However, if planets are not "derivative products" of solar formation the sun could not have transferred angular momentum to the planets. Why they orbit and rotate the way they do around the sun would have to do with their own origins and evolution.

Magnetic braking can not account for the many different orbits and rotations of planets, but this is not inconsistent with transformation hypothesis. Although a planet's distance from the sun determines its orbital velocity, each planet was fully formed when it came to its current orbital position.

Magnetic braking can not explain the retrograde rotations of Venus and Uranus. Lots of moons, such as Triton and Phoebe, also spin in the opposite direction of their host planet. Because of this it has been proposed that some catastrophic event is responsible for these aberrations. These moons, and the Pluto-

Charon system, were freed from the Kuiper Belt and later become unbound gravitationally from the planet Neptune.

Neither camp has an explanation for the mechanism. However, transformation hypothesis considers the solar system as "unstable and subject to change" so these sorts of catastrophic events are not inconsistent with it.

The giant gas planets Jupiter, Saturn, Uranus, and Neptune resemble our sun, so it is assumed they are closest in age to it. While this is consistent, once again, with transformation hypothesis, because of the dominance of nebular hypothesis the larger number of folks in the astronomical community likely still hold that stars and planets are "qualitatively different."

Jupiter, Saturn and Neptune are a source of "radio emissions" with an unknown cause and emit more "energy" then they receive from the sun. Uranus has no such, as of yet, detectable "energy production" but this may be because of its orientation to Voyager II's sensors.

Earth, Venus and moons of Saturn and Uranus also have characteristics similar to stars, such as an "internal energy source." We can consider, in some cases that the activity in the core of planets or moons may be as a result of some unknown external cause, as may be the case with Jupiter's moon Io.

The atmospheres of the gas giants and the sun are similar chemically, and this is consistent with transformation hypothesis as well. The range of hydrogen to helium is close. Brown dwarf stars are also similar to the sun and gas giants in this respect. Never the less, nebular hypothesis says that stars and planets form differently in that brown dwarfs can not get enough mass to sustain nucleosynthesis even though they are a result of "a primary process" whereas gas giants are derivative.

If there isn't enough mass to accrete and produce a gas giant (as required by the proposed limit of 13 Jupiter masses), then why doesn't the accretion process continue and produce a smaller than brown dwarf body? Because that would mean that this object is a gas giant and convention says gas giants are derivative, or "secondary formations," we are told.

The sun and the gas giant have atmospheres which have what is known as differential rotations which has to do with the rotating fluids being "in hydrostatic equilibrium." If these giants formed their rocky cores by accretion and gravitationally attracted elements forming gas atmospheres, what happened to the rocky planets closer to the sun? Current wisdom says the lighter elements were pushed away by the solar wind. Yet gas giants have been recently found orbiting other stars within the minimum distance requirement of nebular hypothesis.

Transformation hypothesis says that gas giants are in an earlier phase of stellar evolution, and later the gas envelopes will be shed revealing the rocky cores within.

Chapter Forty Nine - Brown Dwarfs and Migrating Planets

Planetary Evolution Part Five

The fourth in the series of 7 papers, by Anthony Abruzzo goes into greater detail on brown dwarfs and compares nebular and transformation hypotheses.

It is common for proponents of Nebular Hypothesis to say that Brown Dwarf Stars are the missing link between stars and planets. Anthony says that that is a misnomer as far as derivative hypotheses are concerned, but "is consistent within the general framework of Transformation Hypothesis."

If planets form as a secondary process of accretion, then they are derivative, or "by-products of stellar evolution" and there is no connection between stars and planets. Brown stars represent the end of stellar evolution to the Nebular Hypothesis. However, if the term "missing link" is more appropriately used in the context of Transformation Hypothesis, then it is clear that Brown Dwarfs represent a point between stars and planets and they are just "one link in an evolutionary chain."

From Gas Giants to Rocky Planets

In his fifth paper, Abruzzo goes into greater detail on the transformation of gas giants into rocky planets discussing empirical evidence for rocky cores of our moon and the Earth, composition of the gas envelopes of the gas giants, primary and secondary atmospheres as a function of mass and age, and the external atmosphere dissipation mechanism.

Abruzzo cautions us to be patient as we gather more empirical evidence about our solar system, and suggests that we take a wait and see attitude as New Horizons sends back data on Pluto, answering some questions and raising others. (The article was written prior to New Horizon's arrival near Pluto).

Migration Hypothesis

The 6th paper on Planetary Migration Hypothesis discusses resonances and rightfully points out that there is no theoretical basis for these observations.

Planetary migration depends on the Nebular Hypothesis' planetesimals, and unless planets formed from gas and dust from the proto-planetary disk there could have been no formation or migration possible.

Since the Solar System is said to have all formed from one evolutionary process, the Migration Hypothesis fails, yet Transformation Hypothesis, which sees the Solar System "as a work in progress," can accommodate planetary migration.

Origins of Nebular Hypothesis

The 7th and final paper in the series discusses the origins of the Nebular Hypothesis. Emphasis is made that it is "still just a hypothesis," and that it will be superseded by newer hypotheses "if it can not support the incorporation of new data that proves inconsistent or 'anomalous,' in the Kuhnian sense, with its theoretical architecture."

Clearly, Anthony Abruzzo is talking about the mainstream notion of the scientific method, which conflates hypothesis with theory and relies strictly on empirical evidence and proofs. It is in this light that the author of these papers examines the origin of Nebular Hypothesis and also offers up his critique.

He traces the earliest speculations of a nebular hypothesis to the ancient Greeks and the first of the catastrophic hypotheses to Buffon in 1745. From Anaxagoras of Clazonmenae's "nous" giving rise to the universe in the 5th century BC, to the Epicurian, Lucretius' "On The Nature Of Things" Abruzzo accounts for the evolution in thinking about a "materialistically based evolutionary cosmogony."

The "Aristotelian/Ptolemaic-based cosmology" gave way to Nicholas of Cusa, Giordano Bruno and Rene Descartes' cosmology. The classical Nebular Hypothesis came in the 18th century with Swedenborg, Kant and Laplace. Kant's 1755 "Universal Natural History and Theory of the Heavens: An Essay on the Constitution and Mechanical Origin of the Whole Universe Treated According to Newton's Principles" was superseded by Laplace's "The System of the World" based on Newtonian principles and explaining the 5 phenomena of the Solar System:Planets move in the plane which passes through the center of the sun.

- Planetary satellites move in nearly the same plane as their primaries
- Planets rotate in the same direction as the sun
- Nearly circular orbits of the planets and their satellites
- High eccentricities of cometary orbits

Difficulties with Nebular Hypothesis led back again to a catastrophe based approach with the likes of Thomas Chamberlin, Forest Moulton, James Jean and Harold Jeffreys. These had their own set of problems and fell under the general category of derivation where planets formed from matter originating in the sun.

With derivational hypotheses, all 5 of the Laplace phenomena are held to be a result of a common origin. Abruzzo asks if it can not just as easily be assumed that the characteristic of the planets in the Solar System "merely reflect how the sun's gravitational field influences their motion, not as an originating cause, but, rather, as an ordering cause?"

Just referring to "Solar System" predisposes one to think of a common origin and interpret facts in that light, says Anthony.

Chapter Fifty - Planetary Evolution and the Rational Scientific Method

Planetary Evolution Part 6

Applying the Rational Scientific Method to Transformation Hypothesis in the papers by Anthony Abruzzo, we note the lack of scientific (or any other) definitions for Key Terms. However, Abruzzo is using the same method of inquiry as proponents of the Nebular Hypotheses: observation, proof, evidence, etc. He does not take issue with their terms, only their assumptions and conclusions. We need not consider in great detail all the Key Terms, however we do remind the reader that a scientific definition must NOT be circular, synonymous, ambiguous or contradictory. Also, we assume the ideas to be based on mainstream "understanding" and "definitions" for universe, space, mass, gravity, electricity, magnetism and light, etc.

Transformation Hypothesis simply lays out the general idea that one star forming one planet has less problems than a common origin for the sun and all the planets in the solar "system" described by "derivative hypotheses." As such, we looked only at some of the common objections to Nebular Hypothesis, and the basic assumptions of Transformation Hypothesis.

As pointed out previously, the terms cosmology and cosmogony by definition proclaim a beginning to "The Universe" and suggests, potentially, an end depending on which version of Big Bang one adheres to. Abruzzo points out another unscientific bias when he says that just referring to the Solar "System" predisposes one to think of a common origin and so it follows that they would interpret "facts" in that light.

The Universe, being defined as "all of time and space and its contents," fits nicely with Big Bang cosmology's idea that "all of matter and time and space were once inside a small dimpled pea." Abruzzo opts for an indeterminate age of the universe rather than the absurd Big Bang Theory (BBT) or eternal Steady State Theory (SST) in order to contrast stellar transformation to derivation. It is rather convenient for Transformation Hypothesis to avoid the time constraints of BBT and the "metaphysical" question of eternity. This position also can accommodate the Proton Decay Hypothesis (as Abruzzo sees it) which claims to account for loss of mass.

Yet, we need to suspend our rational thinking in order to account for Big Bang creation of matter, time and space, or Proton Decay hypothesis' annihilation of mass. Both BBT and SST call for the expansion of space, whereas matter density in SST remains the same because of constant creation.

Anthony's claims that neither Steady State nor Big Bang can be proven or scientifically analyzed, and to question whether the universe had a beginning or not falls "outside the purview of science" are specious at best. Matter and motion are eternal, and rationally can not be any other way. Likewise, Proton Decay Hypothesis, where the entire universe will eventually disappear, is also an irrational proposal.

Creation and annihilation of matter, in this context is ridiculous, and deserves no place in a scientific discussion. Big Bang, Steady State, or any other kind of creationism fails on multiple accounts, as revealed in previous articles and chapters in my books on Rational Science, and as discussed with the luxury of detail at Rational Scientific Method Facebook group.

Let's take a cursory look at some of the assumptions of Transformation Hypothesis. The first article by Abruzzo sums up Transformation Hypothesis like this:

- Planets should be viewed as "end products not by-products of stellar evolution."

- Overlapping masses between classes implies "an evolutionary continuum."
- Loss of mass and contraction occurs simultaneously.
- Proton decay could account for the gamma radiation we observe, and also can account for the cosmic microwave background.
- In 1994 terrestrial gamma flashes were detected in the earth's upper atmosphere, and it is supposed that the mechanism for this is proton decay.
- There may be other mechanisms for mass depletion not known at the present.
- Whether or not the universe had a beginning should be left to the metaphysicians.

The first point is an assumption of the Transformation "Hypothesis" and hopefully used to substantiate its author's analysis. I refer loosely to this as a hypothesis and prefer "analysis" over theory, because, although it is similar in form to mainstream's scientific method, it does not conform to the Rational Scientific Method of hypothesis (RSM):

- illustrated objects
- definition of Key Terms
- assumptions

There must first be a hypothesis. Only then can the theory explain the phenomena.

Points 2 through 6 can not be accepted as assumptions until and unless the term mass is defined. Point seven must be dismissed for reasons already stated. We can only guess as to the meaning of the term "mass" since none was provided. Dictionary.com gives their version of this term:

"In physics, the property of matter that measures its resistance to acceleration. Roughly, the mass of an object is a measure of the number of atoms in it. The basic unit of measurement for mass is the kilogram."

Therefore, if we understand mass to be number of atoms, then it would appear Abruzzo is talking about a "quantity of matter" in points, 2, 3 and possibly 6, but his take on Proton Decay Hypothesis refers to the complete annihilation of matter, so we have to ignore points 4 and 5.

If we assume that mass is the number of atoms, then matter is the set of all atoms.

Proton decay, according to other sources, says that protons break down into smaller sub-atomic particles. Which begs the question, what happens to the rest of the "atom?"

None-the-less, enter relativity and we find there are different types of mass; inertial mass, rest mass, and gravitational mass. It is clearly unscientific according to RSM to have multiple meanings for a Key Term in a scientific hypothesis, so we will stick to what seems to be the appropriate definition for mass (a quantity of matter/number of atoms) in the context of planetary evolution. Mass is discussed in great detail in several chapters in the book Rational Science Vol. I and III. (Relativists state that mass increases with velocity. This is rationally impossible since we rule out creation of matter.)

The mainstream scientific method, being what it is, has built layer upon layer of failed hypotheses in arriving at our various positions on cosmology and cosmogony. We've seen that at the heart of our problem is the lack of scientific definitions. But proving equally contrary to our understanding phenomena is not clearly understanding and stating the difference between objects and concepts. Not since our early readers about Dick and Jane, have we been able to distinguish between verbs and nouns, or objects and concepts.

Science, more often than not, defines objects and describes phenomena rather than describing objects and explaining phenomena using those objects. Instead of using illustrations, the real language of science, cosmologists use abstract mathematical concepts in an attempt to explain various phenomena. It places the observer front and center in all its efforts to be objective and resorts to experimentation, proof and evidence which results in quite the opposite: subjectivity.

Take for example, one of the main arguments against Nebular Hypothesis; angular momentum:

Angular momentum is equal to mass times velocity times the distance from the center of mass, and must remain constant in an isolated system. This explains nothing. It only describes a phenomenon.

As a small boy, I went to the Santa Barbara, California NASA facility on a school field trip. During a presentation, a NASA spokesperson had me come up on the stage and stand on a round platform with wheels under it. He had me stick my arms out before spinning me around. Then he had me pull my arms in. When I did this I nearly flew off of the contraption as my rate of rotation increased. This is, in a nutshell, angular momentum.

It does raise the legitimate question, "Why does our sun spin slower than it should based on the predictions of Nebular Hypothesis and the corresponding math? For me, this also raises an even bigger question, "Why don't we use objects to explain phenomena in our hypotheses instead of depending so much on mathematics?"

Abruzzo makes a good point about angular momentum being a problem for Nebular Hypothesis. He does admit that neither camp has an explanation for the mechanism behind various phenomena. Magnetic braking is unable to account for retrograde rotations and planet's radio emissions exceeding the energy received from the sun are also problematic, says he.

The author of the articles about Transformation Hypothesis raises some good questions, most of which were covered in the previous articles so I won't belabor that here.

Abruzzo admits that we have limited "knowledge" of our solar system. He says that we are "restrained by the limitations the empirical data imposes on us" and cautions us to be patient as we gather more empirical evidence about our solar system.

But science is not about knowledge, or evidence, or proof as we can plainly see when we note such things as "the atmospheres of the gas giants and the sun are similar chemically." This is consistent with both Nebular and Transformation Hypotheses. In other words, either proponent can clam this is proof of his or her hypothesis.

The case has not been made, in my opinion, as to whether the Solar system was all formed at the same time from a single origin, nor has Transformation Hypothesis, which sees the Solar System "as a work in progress" settled the matter. Intriguing as it may be, it has raised as many questions as it has answered.

When emphasis is made that it is "still just a hypothesis," and that it will be superseded by newer hypotheses, the weakness of the mainstream scientific method is made evident. The mainstream scientific method conflates hypothesis with theory and relies on subjective empirical evidence and "proof."

The "5 phenomena of the Solar System…"

1. Planets move in the plane which passes through the center of the sun

2. Planetary satellites move in nearly the same plane as their primaries
3. Planets rotate in the same direction as the sun
4. Nearly circular orbits of the planets and their satellites
5. High eccentricities of cometary orbits

admittedly, can have any number of possible mechanisms. Neither creation of the fundamental unit of matter, nor annihilation as in proton decay can be considered.

With derivational hypotheses, all 5 of the Laplace phenomena are said to be a result of a common origin. Abruzzo rightly asks if they may also be assumed, "not as an originating cause, but, rather, as an ordering cause?" The answer is yes.

The answer to how planets and stars are formed may lie somewhere in the middle between the two opposing views. It is rather like unraveling a knot, It is far easier to do so once the two ends have been found, but we are not likely to find an answer by refusing to address critical questions as is suggested by taking the middle ground of "a universe of indeterminate age."

Without applying the Rational Scientific Method we are never going to arrive at any explanations for such things as planetary evolution. Can the Rope Hypothesis offer a model for solar system and galactic formation? Stay tuned!

Join MonkE and the Rational Scientists for a lively discussion:

https://www.facebook.com/groups/RationalScientificMethod/

Other Books by Monk E. Mind

Found on Amazon.com and http://www.bhwservers.com/monkemind

From Short Shorts – A collection of short stories, and really short stories

The End of the Trail

He walked along the trail with all the other workers. They had toiled all day in the field, and now were heading back to join the rest just over the hill. His kind had lived and worked this land for over a thousand years. They were the hardest workers anyone had ever known.

They were all tired and hungry, and it was quiet as they mindlessly shuffled down the trail. He had walked this way many times before, as they all had, without a single thought about the individual sacrifice each had made for the collective. This was the way it had always been. His large strong body moved forward with no thought about what tomorrow would bring. In fact, he didn't think anything at all. None of them did.

Suddenly a bright white intensely hot beam of light shot out of the sky. His legs curled up underneath him as he collapsed, instantly dead. His insides were cooked, and a single puff of smoke rose from his body with a pop.

"Time to eat" Jimmy's mother called from the back porch. Jimmy put his magnifying glass in his pocket, and muttered under his breath, "Stupid ants".

DON'T MISS BUG WORLD ON AMAZON.COM

www.ingramcontent.com/pod-product-compliance
Lightning Source LLC
Chambersburg PA
CBHW071412170526
45165CB00001B/247